Praise for

I AM A PART OF INFINITY

"*I Am a Part of Infinity* is a beautifully written exploration of Einstein's work and his most profound question: What power so deeply connects the human mind with the universe that we can grasp the very laws of nature? This essential work examines the intersections of science, spirituality, and philosophy, offering fresh and meaningful insights into one of history's greatest thinkers and the ideas that shaped his entire worldview."

—David Fideler, PhD, editor of *The Pythagorean Sourcebook and Library* and author of *Breakfast with Seneca*

"By digging deep into archival, epistolary, and biographical material, including Einstein's large personal library, Kieran Fox reveals the paradigmatic scientific genius to be a seeker whose lifelong appreciation of the beauty and mysteriousness of the cosmos made him an ardent believer in a rational universe comprehensible by mathematics. Einstein's unusual spiritual path ultimately led him to become a practitioner of pacifism and vegetarianism and to identify nature with a cosmic consciousness. Fox traces this triptych back to Einstein's extensive readings of the idealist Arthur Schopenhauer, the panpsychist Baruch Spinoza, and the Hindu Upanishads. *I Am a Part of Infinity* is a delight to read at a point in time when the limits of materialism are coming into clear and public view."

—Christof Koch, PhD, author of *Then I Am Myself the World*

"In this meticulously researched book, Kieran Fox shows us a long-ignored and misunderstood side of Einstein: his spiritual conviction that we are entwined with the universe. Fox delves deep to reveal the hidden influences, from Pythagoras to Jainism, that inspired Einstein's epic quest to create a 'cosmic religion' in which mind and matter would be melded into one intricate, all-encompassing whole. *I Am a Part of Infinity* is gripping, inspiring, original—a must-read for anyone who wants to understand the mind of the great physicist."

—Jo Marchant, PhD, author of *The Human Cosmos*

"We have long known that Spinoza was Einstein's favorite philosopher. With Fox's wonderfully readable book, we are finally treated to a clear and accessible explanation of why this was so, and how Einstein's own spirituality was inspired by the 'heretic' of Amsterdam."

—Steven Nadler, PhD, author of *Spinoza*

"Science and spirituality often seem like contradictory approaches to existence. But Kieran Fox shows that, for Einstein, they were in fact complementary and convergent ways to explore reality. In this beautiful elaboration of the deeper levels of Einstein's scientific mind, Fox discovers the unexpected spiritual side of the most celebrated genius of modern times. Wonderfully written and easy to read, *I Am a Part of Infinity* is filled with surprises and 'Wow!' moments awaiting the reader."

—Georg Northoff, MD PhD, author of *Neurowaves*

"Like Isaac Newton's alchemy or Johannes Kepler's astrology, most scholars have seen Einstein's writings on religion as a distraction from his scientific work. Yet Fox makes a bold argument that Einstein's effort to build a modern spirituality around curiosity and awe inspired his own science and also deserves a wider hearing among scientists today. Deeply researched and revelatory, *I Am a Part of Infinity* is essential reading for anyone interested in science and religion."

—Alex Soojung-Kim Pang, PhD, author of *Rest*

"A groundbreaking and gripping account of Einstein as a visionary herald of a new scientific and post-religious spirituality for our times."

—Evan Thompson, PhD, author of *Waking, Dreaming, Being*

I AM
A PART OF
INFINITY

I AM
A PART OF
INFINITY

THE SPIRITUAL JOURNEY OF
ALBERT EINSTEIN

KIERAN FOX

BASIC BOOKS

New York

Basic Books
Hachette Book Group
1290 Avenue of the Americas, New York, NY 10104
www.basicbooks.com

Printed in the United States of America

First Edition: April 2025

Published by Basic Books, an imprint of Hachette Book Group, Inc. The Basic Books name and logo is a registered trademark of the Hachette Book Group.

The Hachette Speakers Bureau provides a wide range of authors for speaking events. To find out more, go to hachettespeakersbureau.com or email HachetteSpeakers@hbgusa.com.

Basic books may be purchased in bulk for business, educational, or promotional use. For more information, please contact your local bookseller or the Hachette Book Group Special Markets Department at special.markets@hbgusa.com.

The publisher is not responsible for websites (or their content) that are not owned by the publisher.

Print book interior design by Amy Quinn.

Library of Congress Control Number: 2024031523

ISBNs: 9781541603578 (hardcover), 9781541603592 (ebook)

LSC-C

Printing 1, 2025

To my dad, for giving me all the right books
to read when I was growing up

And to my mom, for always encouraging
me to write one of my own

A human being is a part of the whole, called by us "Universe," a part limited in time and space. He experiences himself, his thoughts and feelings as something separate from the rest—a kind of optical delusion of his consciousness. The striving to free oneself from this delusion is the one issue of true religion.

—Albert Einstein,
letter to Robert Marcus,
February 12, 1950

CONTENTS

INTRODUCTION

THE SACRED SCIENCE

In this materialistic age of ours the serious scientific workers are the only profoundly religious people.

—**Albert Einstein, "Religion and Science" (1930)**

A LBERT EINSTEIN IS AN ENIGMA.

This is a strange claim to make about the most famous physicist of all time—someone whose name and face are known to all, a man who remains renowned around the world for revolutionizing our understanding of the cosmos. But in essence, Einstein is still an unknown. Although everyone recognizes him as the quintessential quirky genius, very few realize that the celebrated scientist had a deep spiritual side. Even as his phenomenal mind was trying to fathom the farthest reaches of the physical universe, he was also trying to devise a down-to-earth spirituality that saw the divine disseminated everywhere. Einstein felt that one wondrous force was woven through all things everywhere—and this profound sense of the pervasive sacred influenced every aspect of his existence, from his marvelous science to his passionate pacifism. Underlying Einstein's endless efforts to unify physics and unite humanity was a longing to go beyond all apparent duality and show that every person was a part of Infinity.

Einstein dreamed of discarding the age-old religious dogmas and introducing a dynamic new notion of the divine that would mesh well with a modern mentality. And in some ways, we've succeeded. Science and spirituality seem closer than ever. Buddhist monks hop in brain scanners; the Dalai Lama holds regular dialogues with scientific researchers; and mindfulness meditation has become a mainstay of mental health. Everywhere, secular people engage enthusiastically with spiritual practice. But we're still a long way from feeling Einstein's reverence for reality. The revolution he hoped for in our religious feelings remains unfinished. Countless books have chronicled his life and work, but the real story of his radical religious views has yet to be told.

I never expected I would try to tell this tale, but in retrospect the trail isn't hard to trace. All my life I'd been looking for some synthesis of science and spirituality. I started meditating and studying Spinoza when I was in high school, and when I went to college I double-majored in religious studies and neuroscience. For a while I lived in a Buddhist monastery nearby. I would wake before dawn and meditate for hours, then saunter over to school and study the brain during the day. Even then, it seemed like I was studying one and the same thing from two different sides. But neither science nor spirituality seemed to offer any solution to the strange separation of mind and matter.

After college, I went to live with Tibetan monks who trained me in traditional meditation techniques. High up in the Himalayas, I was happier than I'd ever been. Life was simple but splendid. I had experiences so overpowering I still have a hard time understanding them today. And eventually I aspired to ascend even higher into the mountains and maybe become a hermit who'd never be heard from again.

But in all honesty, I'd started to harbor deep doubts about this destination. The path I was following claimed it could answer all questions, explain all things, contain all experience. But too often I was asked to take things on faith. Too often truth had to defer to dogma. And far too often the very real achievements of the rational mind were maligned. I finally realized that following this path would mean forever forsaking both rigorous thought and physical reality. I would have to force myself to forget about ever finding a more holistic vision of existence.

I came down from the mountains determined to find some way to merge science and spirituality, and for years I tried to grasp the marvelous experiences of the human mind with the finest instruments of modern science. While working on my PhD, I used MRI scanners to try to figure out how meditation shapes the structure of the brain. As a neuroscientist at Stanford, I passed electrical currents straight into people's brains to probe the neural correlates of consciousness. And then I made the mad decision to start medical school in my mid-thirties, hoping that I could someday use spiritual practice to help people suffering from the still-mysterious diseases of the mind. By day, I did scientific research and studied medicine; by night, I meditated and read the world's spiritual literature. But despite my best efforts to integrate all my interests, these two paths mostly ran in parallel, and I didn't expect they'd ever converge. A synthesis of science and spirituality seemed like a dream destined to remain unfulfilled.

Until I came upon a little book called *Quantum Questions*. Alone in a small cabin that sits in an enchanted forest on the slopes of the world's largest active volcano, I encountered something extraordinary. It was a collection of what can only be called mystical writings—and yet these sagacious musings came not from Buddhist monks or Christian saints, but from the founders of quantum physics. Here were paragons of rationality proclaiming that "the personal self equals the omnipresent, all-comprehending eternal self" and insisting that "it was not by any accident that the greatest thinkers of all ages were also deeply religious souls."[1] Here were hardheaded scientists asserting that "every advance in knowledge brings us face to face with the mystery of our own being" and talking about "the translucence of the eternal splendor of the 'one' through the material phenomenon."[2]

These words were written by the greatest minds of the twentieth century—maybe some of the best brains in history. They were masters of mathematics who fashioned the most rigorous and far-reaching models of reality ever made. And I realized that every one of them harbored deep religious feelings; all of them were familiar with and fond of Eastern spirituality; and many of them even made the trip by steamship to visit India, China, and Japan in person. By the time I closed the book, I'd come to a startling conclusion: the people who laid the foundation for the entire imposing edifice of modern physics were not just outstanding scientists, but earnest spiritual seekers.

I soon discovered that the same was true of the sages who'd started the Scientific Revolution some five centuries ago. The mythical minds who made the modern age—Copernicus and Kepler, Galileo and Newton— also had a spiritual side that not only informed their science, but was almost inseparable from it. For them, to fully grasp Nature's laws was to fathom the mind of God. And even five hundred years ago, this feeling was nothing new. The scientists of the sixteenth century realized that they hadn't really revealed a new path at all. Rather, they'd rediscovered an ancient road to reality that had been forgotten for thousands of years. They walked farther along it than anyone ever had before, but all the while they were well aware of their debt to the ancient wayfinders. They knew they were following in the footsteps of one who'd lived long ago, a legendary mystic who taught that math was a mystical path leading directly to the divine mind.[3]

These are the truths the textbooks never talk about: the spiritual convictions behind the scientific creativity, the mysterious motivations that drive certain minds to seek a transcendental pattern pervading all the transient appearances. And in realizing that a religious yearning had always permeated the scientific quest, I was reminded of an old parable told by the Persian poet Rumi. In this archetypal tale, a seeker dreams of a great treasure in some faraway land and sets off on a journey to find it. But after countless years of searching, the quest yields nothing except the knowledge that the treasure had been hidden in his own home all along, buried beneath his feet. In the end, only by returning to his roots does he find what he's been seeking for so long.[4] I'd spent years searching the world for a spirituality that didn't force me to surrender my common sense, that could at least be *compatible* with science. And yet all the while, at the origin of Western civilization was a system where science and spirituality were the closest of companions—a transformative teaching known as the *hieros logos*: the sacred science.[5]

Albert Einstein was an ardent apostle of this sacred science. And not only did he embrace this ancient tradition: he amended it for the modern era, articulating a spiritual system that honored the great minds of antiquity without ignoring the enormous advances in understanding of our own age. Einstein was eager to share his convictions, so he spelled out his spirituality in simple words in some of the world's most prominent publications.[6] But despite his immense fame and his unparalleled intellectual achievements, his spirituality was mocked and misunderstood during his lifetime.[7] And since his death, it's suffered a fate far worse: the spiritual side of history's greatest physicist has been all but forgotten. Why?

Maybe it's been ignored because it's incomprehensible to mere mortals. It's easy to assume that Einstein's philosophy, like his physics, is simply beyond the grasp of most people. But the truth is much more terrible. It's not that Einstein can't be understood; it's that he asks too much. Just as his science forced physicists to radically revise their most basic assumptions about the fabric of reality, his spirituality challenges us to reconsider our fundamental assumptions about both the form and function of religion. And although many of us are now comfortable with criticisms of religious creeds, Einstein's doctrine demands much more: it also compels us to confront the sublime spiritual feelings that animate the scientific enterprise. Einstein insists on an integration of the apparently irreconcilable, a reunion of reason and religiosity he called "cosmic religious feeling."[8] It takes great imagination to appreciate Einstein's audacious worldview; its luminous unity is a vision most of us are simply not ready to see. But comprehending the cosmic religion doesn't require any fervent faith or blind belief, and it definitely doesn't demand an unusual intelligence.

Not that understanding Einstein is easy, exactly. Really grasping his cosmic religion means getting acquainted with the radical geniuses who anticipated and inspired it: Pythagoras, Giordano Bruno, and Baruch Spinoza in the West; Lao Tzu, Buddha, and the authors of the Upanishads in the East. This is not light reading. But the real difficulty isn't understanding what they wrote; it's *recognizing what they were*. Because these people were not just thinkers or philosophers, not just poets or proto-scientists. All of them were seekers after truth, aspiring initiates of the Infinite. They expected more than just answers to questions; they wanted communion with the cosmos, an alteration of their innermost essence.

Anyone who hopes to approach these teachings must open themselves to the same experience. It's so easy to believe that there's nothing more to us than our fragile little egos, enduring for only a moment in endless time. Cynicism, skepticism, and simplemindedness all conspire to convince us that this is the case. But the core conviction of the cosmic religion, and all the analogous systems that came before, is that consciousness can become far more comprehensive. The human mind can be molded into a mighty instrument, a mirror of the Infinite. "There comes a point where the mind takes a leap," Einstein once said, "and comes out upon a higher plane of knowledge."[9] And he knew from his own experience that everything of enduring value had originated in this higher realm and then been brought back by beings he called "the bearers of a higher level of consciousness." All through history, "the great artists, ethical pioneers, and thinkers" had stormed Heaven and thereby helped "raise human society to a higher level of experience, vision, ethical being, and understanding."[10] So it was obvious to Einstein that we should hold "the highest stage of consciousness as the highest ideal."[11] It was our duty and our destiny to develop the mind so we could keep discovering the endless gifts offered by the inexhaustible origin of all things.

Einstein called the eternal energy underlying all things the "arch-force," and for him this was not some New Age nonsense or some abstract theoretical entity. The arch-force was always with us, evident everywhere, the foremost fact of physical existence. Einstein himself was the first to provide scientific proof of this spiritual principle. With his immortal equation $E = mc^2$, he'd proven that a mighty power permeated all things. An immense energy was locked up inside every atom, only waiting to be unleashed. And it was in Einstein's time that humankind finally tapped the tremendous potential of this transcendental force. Our oldest wish was granted: we became masters of what he called the "basic power of the universe" and were given godlike powers.[12] And the moment this higher force was harnessed, it was welded into a horrendous weapon and used to annihilate hundreds of thousands of human beings.

Einstein spent the rest of his life repenting for having "participated in opening this Pandora's box."[13] In our desperate quest to understand reality, we'd never really asked ourselves if its deepest source had a dark side. But now it was obvious that the omnipresent energy of the Infinite was utterly indifferent. Our new power over Nature merely meant a multiplication of our potential for creation *and* destruction, demonic and divine. Would we ever be worthy of wielding these awesome powers? Was it "possible," Einstein wondered, "to control man's mental evolution so as to make him proof against the psychosis of hate and destructiveness?"[14]

In ancient Greek mythology, after Pandora accidentally allows all the evils to escape into the world, one last power still remains at the bottom of the box: hope.[15] And Einstein likewise held that at least one ray of light remained. Just as a "higher level of consciousness" had magnified our potential for both good and evil, the same exalted experience could liberate us "from the bondage of egocentric cravings, desires, and fears" and give us the wisdom to use this wondrous power wisely.[16] "As man becomes conscious of stupendous laws that govern the universe in perfect harmony," Einstein explained, "he begins to realize how small he is. He sees the pettiness of human existence, with its ambitions and intrigues, its 'I am better than thou' creed. This is the beginning of cosmic religion within him. Fellowship and human service become his moral code."[17]

Make no mistake: Einstein was calling for nothing less than a new kind of consciousness. He was asking us to expand our minds until our ethical enlightenment equaled our intellectual ingenuity, to initiate an inner revolution just as radical as the transformation he'd helped inaugurate in the scientific sphere. "Often in evolutionary processes a species must adapt to new conditions in order to survive," he noticed. And with our mastery of matter and our new power over Nature, he was convinced that "the human race consequently finds itself in a new habitat to which it must adapt its thinking."[18] The key to continued evolution was what Einstein called "the highest kind of religious feeling."[19] A truly conscious mind could see that we were simply small nodes in an infinite network of being where all bodies and all boundaries were only conventions. And Einstein felt we needed to mirror this underlying metaphysical unity in thought, word, and deed. Every one of us had to embrace an ethics inspired by interconnectedness and aspire to become "whole" human beings, "made in the image and likeness of the arch-force."[20]

Yet imitating the Infinite didn't mean dissolving into the divine and drifting off into a dreamy detachment or indifference. By no means did Einstein believe we should go beyond the body or escape from earthly existence. On the contrary, he contended that "no one has the right to withdraw from the world of action at a time when civilization faces its supreme test."[21] Einstein's challenge to each and every one of us was to channel the prodigious creative power that permeated all things and put it to use for pure purposes. "We cannot stand aside and let God do it," he insisted. "Whatever there is of God and goodness in the universe, it must work itself out and express itself through us."[22] This was always Einstein's ideal: not an otherworldly dreamer lost in "cloud-cuckoo land," but an active participant in the Infinite, "an idealist" who "lived on earth."[23]

So perhaps it's only fitting that this book was written not in a monastery or on a mountaintop, but while I was in the middle of medical school. My days were spent in the hospital, surrounded by all the gruesome spectacles you'd expect; my nights were spent studying the maddeningly complex molecular biology that forms the basis of modern medicine. Every moment that remained went into making the work you now hold in your hands. Writing this book required an enormous amount of research and reading (I now have an entire bookshelf dedicated to the topic). It was the greatest intellectual effort I've ever made. But as I immersed myself in the high philosophy of some of history's greatest thinkers, I realized that the essence of their teaching could never be understood by the intellect alone. A real understanding of pantheism requires a radical alteration of perception.

It wasn't easy to see anything divine amidst all the death and disease that defined my daily grind. But whenever I could, I wandered off into nature, and alone in the wilderness I entered Infinity and felt the arch-force filling all things. It was the solid, silent foundation of every stone; the dynamic energy flowing through the rivers and blowing in the breeze; the inexhaustible force finding expression in every blooming flower. At times, these feelings were so strong I couldn't believe they would ever end. I was actually afraid I could never return to my regular life. I would have to quit doing science, stop studying medicine, and just retire to some secluded spot to revere the reality all around me.

But inevitably, the light of nature would be dimmed by the demands of day-to-day life. Regular reality would reassert itself, and I was able to keep caring for my patients, conducting my research, and completing this book. Once seen, however, that light never fades away entirely. And although its radiant essence can never be captured, in a sense it can be concentrated: its rays can be focused to form an imperfect image of the eternal and unimaginable. Such faint reflections provide the feeble light that illuminates this work. It is the intellectual afterglow of a joyous spiritual experience. It's no exaggeration at all to say that writing this book changed me; I hope that reading it will change you, too.

1

AN UNFINISHED QUEST

I shall conduct the reader over the road that I have myself travelled, rather a rough and winding road, because otherwise I cannot hope that he will take much interest in the result at the end of the journey.

—Albert Einstein, "Cosmological Considerations on the General Theory of Relativity" (1917)

I N THE EARLY MORNING HOURS OF APRIL 18, 1955, PERHAPS THE GREAT-est genius who ever lived passed away in Princeton, New Jersey. The day before he died, Albert Einstein made his peace with death. "I have done my share," he declared. "It is time to go."[1] And although he'd no doubt earned the right to rest, he expected no reward. Einstein never believed in an afterlife, "although feeble souls," he said, "harbor such thoughts through fear or ridiculous egotism."[2] Another form of immortality awaited instead. Already a legend in his own lifetime, renowned around the world for revolutionizing physics, Einstein was leaving behind a magnificent mental legacy that would live on long after he was gone.

Still, he knew better than anyone that his work was unfinished. Not content with the countless contributions he'd made before forty, he spent his final decades obsessing over a theory that would unite all of physics, from the infinitesimal electron to the immensity of entire galaxies. For Einstein, this search for unity was "the highest and most sacred duty," and he vowed that he'd "never give up the hope that this greatest of all aims can really be attained."[3]

It wasn't just the unification of physics he hoped to achieve in his lifetime. Just as ambitious—maybe more so—was his wish that humanity would be unified in a peaceful global civilization. Einstein had been an outspoken pacifist all his life, and he didn't forget his ideals as the end approached. Just four days before he died, he signed a peace manifesto urging people to set aside personal beliefs and political bias. "Remember your humanity," it read, "and forget the rest."[4] And on the table by his deathbed lay the handwritten draft of a speech in which he called for "universal ideals of peace."[5] The quest to unify physics occupied Einstein's final decades, but the promotion of peace occupied his final days—even his final hours.

Everyone acknowledges that Einstein was a first-rate physicist and a passionate pacifist. But what almost no one knows is that in his mind, unifying physics and uniting humanity were simply two different aspects of a single spiritual quest. For Einstein, both the fragmented laws of physical nature and our seeming separateness from one another were just an "optical delusion" of our limited human minds. "The striving to free oneself from this delusion," he said, "is the one issue of true religion."[6] This true religion required not just new beliefs, but "a new type of thinking" to help the human mind "move to higher levels."[7] No longer imprisoned by individuality, the new kind of consciousness Einstein envisioned could confidently pronounce itself "a part of Infinity."[8] And although he knew it was "very difficult to elucidate this feeling to anyone who is entirely without it," he believed it to be the birthright of every human being on Earth to experience themselves as an element of the eternal.[9]

Yet today, what Einstein called "cosmic religion" is almost completely unknown. Like his incomplete unified field theory in physics and his still-unfulfilled hopes for world peace, the renaissance he hoped for in the religious realm remains unfinished. Although Einstein believed that human beings were "hungry for spiritual nourishment," he also accused his age of being "barbarous, materialistic, and superficial."[10] A century later, we're still in a similar predicament. In our increasingly secular and cynical age, many of us search the far corners of the world for some kind of spirituality without superstition, a sense of the sacred without the supernatural. We yearn to satisfy our intrinsic religious impulses without sacrificing our reason or integrity—or our creature comforts. And all the while, Einstein's teaching lies buried at our feet, forgotten. But what was this cosmic religion that Einstein hoped to see embraced by the whole world—and what might it offer us today?

The spirituality Einstein envisioned was so radical that he dismissed every one of the world's existing religions as beyond redemption. But his vision was also so inclusive that Catholic saints like Francis of Assisi, ancient atheists like Democritus of Abdera, modern heretics like Baruch Spinoza, and Eastern sages like Gautama Buddha could all be counted among its enlightened exemplars. Neither a nostalgic revival of Western faith nor a naïve imitation of Eastern philosophy, Einstein's spirituality was something more: an almost alchemical amalgam of mind and matter, a novel synthesis of noble spirituality and exquisite science. It was not so much a system or a specific set of beliefs as it was a heuristic that could help us see that all things emanated from a single sacred source.

The dream of achieving a nondualistic understanding of existence has always been an uncommon ambition, a path pursued by an exceptional few. In stark contrast, mainstream religions have always been enamored with dualism: all of them seem determined to maintain the division between body and soul, above and below, Heaven and Earth. Science hasn't done much better. Early scientific thinkers like René Descartes were just as dualistic,[11] and today there's a tendency to deny the existence of consciousness altogether,[12] or to dismiss the mystery of the mind as forever beyond science's proper domain.[13]

Einstein demanded more. The cosmic religion called for a quantum leap in consciousness, for a full and final recognition that the physical reality studied by science and the fantastic realms explored by the spirit were really one and the same. Until we recognized that "physics and psychology are only different attempts to link our experiences together by way of systematic thought," we'd be condemned to keep living within the confines of the old dichotomies.[14] Lost in dualistic delusions, we'd keep forgetting that matter was only the medium for our minds, that our actions in the outer world were only an expression of our inner essence, that the technologies fashioned by our science only actualized the dreams—and nightmares—born deep in our souls.

Einstein's insistence on an integration of science and spirituality perplexed his friends and exasperated his enemies, but he himself saw no contradiction. "All religions, arts and sciences are branches of the same tree," he said. This "essential unity" had been mostly forgotten in recent centuries—so much so that the two sides now regarded each other with "senseless hostility."[15] Yet in spite of this schism, Einstein still saw "strong reciprocal relationships" between "the realms of religion and science."[16] Just as he felt that "scientific theory brings into play the higher spiritual faculties," he also insisted that "true religion has been ennobled and made more profound by scientific knowledge."[17] For all the apparent incompatibilities, he was convinced that "in truth a legitimate conflict between religion and science cannot exist."[18]

Einstein condensed this conviction into one of his most famous turns of phrase: "science without religion is lame; religion without science is blind."[19] It was a radical claim in Einstein's day. But in hindsight, we can see that he was simply a hundred years ahead of his time, a harbinger of those who now call themselves "spiritual but not religious."[20] These are people who accept the validity of objective scientific knowledge, yet also acknowledge the value of subjective spiritual experience. The goal of these modern seekers is simple: "to be the student and beneficiary of all traditions, and the slave to none."[21] Einstein couldn't have said it better himself. "To see with one's own eyes, to feel and judge without succumbing to the suggestive power of the fashion of the day"—this "freedom of the spirit," this independence of mind, was always his ideal.[22]

Our spiritual independence had come at a cost, of course: what Einstein called "the religious paradise" had been "lost" forever.[23] It's all too easy to blame science for our expulsion from Eden, to resent the rational mind for depriving us of the delusions that have kept us comfortable for millennia. And it's not only religious believers who lament our lost innocence. Nowadays, even some scientists mourn for the old myths that once gave our cosmos meaning. "The more the universe seems comprehensible," says Nobel Prize–winning physicist Steven Weinberg, "the more it also seems pointless."[24]

And it's hard to deny that there's something almost monstrous in the modern view of the universe. Our cosmos seems extravagant in the extreme, profligate almost to the point of being obscene. Billions of stars are born from the basic material matrix and burn so very bright, only to end in ashes. Countless living beings well up from the fertile womb of the physical world, only to wither away again. All things, even the most intricate expressions of the arch-force, are destined to turn to dust. Nowhere is there any evidence for the comforting Creator of our old religious traditions, the good Lord who kindly arranges all things in accordance with a well-considered plan. Faced with the immensity and incomprehensibility of existence, it's tempting to believe that maybe ignorance is bliss after all. Science has stripped us of our old superstitions, and for many of us nothing has been able to fill the void in our souls. *"Nostra culpa!"* Einstein admitted.[25] Our fault.

And yet in Einstein's eyes, science had done much more than just banish the old gods. It also offered us "a new revelation," an astonishing origin story to replace the "fairytales about creation" clung to by the old religions.[26] And as it turns out, the new mythos made by the rational mind meshes surprisingly well with what spiritual seers have been saying for centuries: *all things share a single origin.* Despite all the apparent diversity of our present existence, cosmologists have been forced to conclude that everything was united at a single center in the distant past, an infinitesimal point of pure physical potential they call the primordial singularity.

Billions of years ago, the primordial singularity exploded in the Big Bang. The sublime little seed sprouted and grew to gigantic proportions, giving rise to trillions of glimmering galaxies, an entire glorious cosmos. And although the One multiplied itself into Many, disintegration didn't mean destruction. Everything around us and inside us, our entire exuberant existence, *is* the primordial singularity—no longer simple and singular now, but undergoing an unfolding so marvelous it makes a mockery of all the models of the human mind.

For Einstein, the flowering of our physical world would forever remain mysterious. It was not for us to grasp the reasons, if any, for reality's metamorphosis from integrated monad into infinite multiplicity.[27] But for all its mystery, Einstein believed this to be "the most beautiful and satisfactory explanation of creation."[28] Science showed us that we were not strangers to the cosmos, not prisoners locked in the material realm as punishment, but rather the rightful residents of a remarkable reality worthy of reverence. "The picture of the physical universe presented to us by the theory of modern science," as Einstein put it, "is like a great painting or a great piece of music that calls forth the contemplative spirit, which is so marked a characteristic of religious and artistic yearning."[29] And it's no secret that he often experienced ecstasy in contemplating what he called "the order, the harmony, the magnificence of creation!"[30]

But there was more to Einstein's spirituality than just a renewed reverence for physical reality. The cosmic religion also required a revolution in our conception of *ourselves*. "During the youthful period of mankind's spiritual evolution," Einstein explained, "human fantasy created gods in man's own image."[31] But now it was time to forget these old fantasies and reimagine our *own* essence: no longer to conjure up gods who resembled us, but instead to recognize that we reflected the resplendent energy immanent everywhere in the cosmos. For Einstein, we were not exiles from Eden or fallen angels forever seeking God's forgiveness. We were emissaries of the Infinite, the progeny of the primordial singularity— newcomers to the cosmos, perhaps, but undoubtedly some of Nature's most promising offspring. In us, the arch-force had organized itself to such an astonishing extent that it could understand its own order, ponder its own origins, and expend its immense energies in any imaginable direction. The mandate and the meaning of the cosmic religion was to make every person "more aware" of their "dignity as a cosmic being"—to remind us again and again that we embodied Infinity.[32]

As ethereal as all this might sound, in essence Einstein's spirituality was straightforward. The cosmic religion required neither a fervent faith nor any abstract intellectual acrobatics. "It isn't important whether people understand this or that philosophical system," Einstein said. "What they should understand is that they are endowed with a mind that has the power to unveil the mystery of life."[33] Einstein's intention was to reveal this inner Infinity to as many people as possible—and he thought he knew how. In a conversation with the poet and pacifist William Hermanns in 1943, he summed up his plan in the simplest possible terms: "We must found a cosmic religion."[34] With the Second World War raging, Einstein was looking for a way to end human conflict forever, and he was able to spell out his endgame in a few simple words. "To me," he said, "cosmic religion means one humanity, one love, one peace."[35] The man who unified matter and energy, space and time, also wanted to unify all hearts and minds. And his hope was that "a new age of peace will be inaugurated when all people profess a cosmic religion."[36]

Einstein was eager to disseminate this new doctrine, but he always despised dogmatism.[37] He insisted that he was "against all organized religion," and the last thing he wanted was to form a cult around himself or his ideas.[38] Hence he founded no school and hoped for no followers. Shortly before he died, he even arranged for his body to be cremated so that his house could "never become a place of pilgrimage where the pilgrims come to look at the bones of the saint."[39] He had no interest in either mindless imitation or mere adulation.

Instead, he hoped to inspire others to adopt his ideals by setting an example. He was acutely aware that "we will not change the hearts of other men by mechanisms, but by changing *our* hearts and speaking bravely."[40] Ultimately, he thought, "the example of great and pure individuals is the only thing that can lead us to noble thoughts and deeds."[41] By embodying his own principles, he hoped he could serve as a kind of catalyst that would accelerate positive change in other people's consciousness. And since this idea of illustrious examples for others to follow was such an important element of Einstein's spiritual vision, it's only fair for us to ask a candid question: Did he live up to his own ideals? Was Einstein an example worth emulating?

From Mother Teresa to Martin Luther King Jr. to Mahatma Gandhi, it's become fashionable to find fault with even the wisest, most magnanimous mentors of humankind.[42] Fair enough. In the name of historical accuracy and intellectual integrity, it's a legitimate task to humanize our highest idols by highlighting their flaws, large and small. So if it's starting to sound like I'll be ignoring Einstein's many faults and promoting him as a prophet for our times, perish the thought. Einstein was no saint. He was a mediocre husband to two wives and a subpar father to two sons; his private letters reveal many marital infidelities;[43] and scholars have recently discovered repugnant comments on race in his early travel diaries.[44] All these failings need to be acknowledged, and no doubt Einstein would have approved of a less fawning look at his legacy. If there was anything he despised more than the dogmatism of institutions, it was the deification of individuals.[45] "Everyone should be respected," he once wrote, "but no one idolized."[46] He knew we needed role models, but he also insisted we be wary of worshipping anyone, no matter how worthy.

In a way, though, Einstein's glaring imperfections *add* to his importance as an exemplar, rather than detract from it. Religions usually insist on seeing their saints as perfect people—infallible vessels through which divine wisdom is communicated to us mere mortals.[47] Einstein shows us that someone all too human can elevate us all the same. In any case, my aim here is neither to raise an altar to Einstein nor to pull him down from his pedestal. My goal—much more humble than either hagiography or humanization—is simply to learn what his spirituality was really all about, and, with a little luck, learn something from it, too. With Einstein, as with every other great figure in human history, admiration doesn't have to mean idolization. We can honor Einstein's virtues without ignoring his vices.

Still, it's not enough to simply acknowledge Einstein's shadow. We also need to make sure the light is legitimate. If Einstein's spirituality is to have any enduring value, it needs to do more than transcend the prejudices of yesterday; it should also speak to us today. And ideally, it should still have things to teach us tomorrow, too. His cosmic religion mustn't reveal itself to be some foolish fantasy in the harsh light of hindsight. Nearly a century after it was first formulated, it must survive the scrutiny of a more cynical age.

Very few thinkers can withstand such a withering gaze. Einstein's wild ambition to save the world must strike many modern readers as hopelessly naïve, and in fact, a hopeless dreamer is exactly how many of his contemporaries characterized him. Carl Jung accused him of "sentimental idealism with shallow enlightenment."[48] Pacifist Romain Rolland, a Nobel laureate for literature, complained that "his genius is limited to science. In other matters he is a fool."[49] And maybe no one said it better, or with more brevity, than J. Robert Oppenheimer: "Einstein is completely cuckoo."[50] This reputation for romanticism recalls a tale told by Plato about the first philosopher, Thales, who fell into a well while walking along looking up at the stars. A clever handmaid is supposed to have said that "he was so eager to know what was going on in heaven that he could not see what was before his feet."[51] An analogous image of Einstein as brilliant but bumbling seems to have been indelibly imprinted on the popular imagination.

Given all the ornery gossip that's emanated from so many great minds, we'd be wise to convince ourselves that Einstein wasn't just a die-hard dreamer out of his depth. And although it really shouldn't need to be said, it seems that it does: Einstein was no fool. Unlike Thales, even when looking up at the heavens, Einstein always kept an eye on Earth. In a sense, he had no choice. The tragedies of his time (the Great Depression, two world wars, the Holocaust) left no one unscathed—least of all a German Jew.[52] Historical calamities aside, his personal life presented its fair share of trials, too. He never even laid eyes on his first child, a daughter who died in infancy.[53] His son Eduard was stricken by schizophrenia and spent most of his life in psychiatric institutions.[54] Two of his cousins were killed in Nazi concentration camps.[55] And Einstein himself was hounded by anti-Semitism, attempted assassinations, and the animosity of his intellectual inferiors all his life.[56] Not that suffering necessarily imparts wisdom, but the ordeals Einstein endured—as a refugee, a father, and a Jew—are a helpful reminder that he wasn't always insulated in Princeton's ivory tower.

And just as he was no stranger to the harsh realities of earthly existence, Einstein was equally at home in the higher realms of the human spirit. This was not just an accident, or the inevitable outcome of being born a genius. Einstein actually went out of his way to expose himself to a bewildering variety of ideas and ideologies. In an era when world travel was possible only via slow-moving steamships, he visited as many countries, and immersed himself in as many cultures, as he could (ultimately he set foot on every continent except Antarctica).[57] His mind also ranged far and wide. Beyond the boundaries of his own field, physics, he cultivated personal relationships with many of the greatest minds of the twentieth century: famous philosophers like Bertrand Russell, explorers of the psyche like Carl Jung and Sigmund Freud, and spiritual sages like Mahatma Gandhi and Rabindranath Tagore.[58]

And then there were the books. Where personal experience was impossible or the opportunities afforded by the present era proved inadequate, Einstein delved into the crystallized wisdom of the past, the insights preserved for posterity by the written word. Even from a young age, he read widely and voraciously. He tackled Kant's *Critique of Pure Reason* at the tender age of thirteen, and he never outgrew his love of deep, difficult reading material.[59] Helen Dukas, his personal secretary for decades, complained that there were "books everywhere" in his home. "I have often wished that Gutenberg had never lived!" she once lamented in a letter.[60] At the time of his death, Einstein's personal library totaled twenty-four hundred books. Very few focused on physics. Instead, his ample collection was packed with great works of philosophy, literature, and religion.[61] Not that book learning guarantees any wisdom when it comes to the ways of the world (any more than suffering does). Still, the most revealing book of all suggests otherwise. Cervantes's *Don Quixote* sat on Einstein's night table, and according to a close friend it was "the book which he enjoys most."[62]

Any man who lost a child, escaped the Nazis, and slept with Cervantes on his bedside table can hardly be accused of a genuinely naïve idealism. Einstein knew that an adherence to noble ideals and a passionate pursuit of grand goals would seem silly to more cynical minds. He didn't care. "Ever since childhood," he once wrote, "I have scorned the commonplace limits so often set upon human ambition."[63] He even admitted that "I'd also be laughing at the crazy Don Quixote, if he didn't also happen to be me."[64] Like Don Quixote, Einstein *chose* to pursue a higher path; unlike Quixote, of course, his eyes were wide open to the many pitfalls such a pursuit entailed.[65]

Einstein was fifty years old when he first formulated his religious philosophy.[66] The cosmic religion is not the idealistic outburst of an innocent youth, but rather the remarkable synopsis of a restless spirit who'd spent decades reading and reflecting, searching and suffering. With an elegant simplicity, Einstein's spirituality integrates the many interests of a versatile mind and the varied experiences of an extraordinary life. Whether we agree with its message or not, the cosmic religion represents the concentrated wisdom of a well-examined life. And even if it strikes our more cynical minds as wildly optimistic or hopelessly naïve, we need to accept that Einstein embraced these ideals intentionally, with full awareness. In his science, his ethics, and even his spirituality, Einstein was quixotic not by nature or naïveté, but by conscious choice. And when a younger generation sought his advice, he advocated the same idealistic approach. "One should not pursue goals that are easily achieved," he advised. "One must develop an instinct for what one can just barely achieve through one's greatest efforts."[67]

Since Einstein always aimed high, we shouldn't really be surprised that he was trying to formulate a religion for the future rather than legitimize the religions of the past. Even at first glance, Einstein gives every indication that we're dealing with something *different*, a system that doesn't dovetail well with the old doctrines and dogmas. A closer look only confirms that the cosmic religion defies any simple definition and doesn't fit well with any of our familiar traditions.

But this is precisely what most people miss, and so the cosmic religion has been misunderstood from day one. Almost everyone blames Einstein for the confusion. Ever since he dubbed himself "a deeply religious non-believer," he's been maligned for his "maddening tendency to be purposefully gnomic or oblique."[68] Others have accused him of being "notoriously ambiguous."[69] And the baffled cardinal of Boston, sensing "the ghastly apparition of atheism" concealed under the cloak of the cosmic religion, confessed, "I very seriously doubt that Einstein himself really knows what he is driving at."[70]

But Einstein's alleged ambiguity isn't the issue. The real problem is that his ideas have always been approached in the wrong way. One of the first things I learned when I started studying psychology in college was the theory that there are two basic ways of learning new knowledge and absorbing new experiences: *assimilation* and *accommodation*. Assimilation is the simpler of the two. With assimilation, we're on familiar ground; new information can be comfortably classified according to old categories, without challenging any of our core concepts. It's the easiest form of mental expansion, the most painless kind of growth possible. In mathematical terms, assimilation is simple addition: two plus two makes four.

Accommodation, on the other hand, is more like integral calculus. Accommodation is what's needed when we're confronted with information or experiences so novel that they require a revision of our old mental schemas. Information isn't simply added to an existing repository of knowledge; our minds must actually *expand* in order to *understand*.[71]

Everyone who's ever studied Einstein's spirituality has aimed at assimilation. For almost a century, scholars have been trying to cram the cosmic religion into existing cognitive constructs. The famous biologist Richard Dawkins thinks that "Einstein was, in every realistic sense of the word, an *atheist*."[72] Walter Isaacson, the bestselling biographer, is instead convinced that Einstein was a *Deist*: someone who believes in a Creator

God, but doesn't believe He's in the miracle business.[73] For Harvard professor Louis Menand, Einstein was simply a *realist*: his cosmic religion "has nothing to do with morality or free will or sin or redemption. It's just a recognition of the way things ultimately are, which is what Einstein meant by 'God.'"[74] And historian of science Alberto Martínez disagrees with them all, instead concluding that "good old Einstein was *agnostic*."[75]

But like trying to force the proverbial square peg into a round hole, or capture the Infinite in the minuscule container we call the human mind, the cosmic religion just won't fit into these familiar categories. Understanding Einstein's spirituality requires something more: an expansion of our emotional range, an openness to unexpected inner experiences, a blurring of the boundaries of the self. And if we really wish to grasp Einstein's unfamiliar gospel, the growth must go on until we know in our innermost depths what Einstein knew in his.

Naturally, this is not a process that everyone is willing, or able, to undergo—and we can decide, if we like, to remain in the dark. But we can't claim a lack of light any longer. In the decades since Einstein's death, the full force of academic scholarship has been brought to bear on the greatest scientist of the twentieth century. Close friends have recalled his thoughts on countless subjects. We have hundreds of his letters and many of his diaries. We know who his favorite philosophers were, and we know the spiritual teachers, living and dead, he most admired. We even know what books he read. Perhaps never in history has so much material been so readily available for us to fathom the mind of a genius. And whatever the result of our research, we'd be wise to remember Einstein's conviction that "the search for truth is more precious than its possession."[76] Even if we discover only a few tiny diamonds amidst all the dirt, every iota of accurate information is important when it comes to a genius of Einstein's caliber. We need to know what he really thought about the religions most of us still follow today—and what he thought ought to replace them.

Finding answers, though, means sailing out into what will feel, for many, like unfamiliar seas. The far islands of the mind inhabited by the pacifists, pantheists, and Pythagoreans who inspired Einstein's spirituality are seldom visited today. But scattered throughout Einstein's many musings on mind and spirit are all the clues we need. For those with the verve for the voyage, we now know the way. Whether we want to go there—and whether we'll like what we find when we arrive—is up to us.

Such a voyage into the intellectual unknown can seem overwhelming. But ultimately every worldview, no matter how quirky or complex, is formed from a few simple elements, and Einstein's spiritual landscape is no exception. By looking at what he wrote, what he said, and even what he read, we can explore his outlook on the five questions fundamental to any philosophy: what he *valued* as beautiful in the world (aesthetics), how he thought we could *know* the world (epistemology), what he thought the world *was* (ontology), how we ought to *act* within it (ethics), and how we might *transcend* our worldly limitations (soteriology).[77]

Seven decades after Einstein's death, not a single one of these questions has been adequately answered. While many books have done an admirable job of untangling Einstein's personal life and explaining his scientific work, studies of his spiritual side remain shallow, sycophantic, and sometimes simply ludicrous. Bewildered biographers tend to bypass Einstein's mystifying religious musings, keeping things short, sweet, and superficial.[78] Meanwhile, theologians who understand the impious implications all too well deliberately distort his views, resulting in ridiculous attempts to equate Einstein with Jesus or to reach a "rapprochement" between Einstein and Thomas Aquinas.[79] More astute apologists for Christianity, like Oxford theologian Alister McGrath, acknowledge that "Einstein clearly wasn't a Christian" but reassure us that his "reading of the natural world" still "resonates or chimes in with the Christian faith."[80]

The only work that can claim to be a serious study of Einstein's spirituality is *Einstein and Religion*, by the physicist Max Jammer.[81] And while Jammer deserves credit for cobbling together many scattered sources that speak to Einstein's spiritual side, he distorts or downplays every important influence and dwells on much that's irrelevant.[82] In the first few pages, he suggests that "Einstein revealed only a small part of what God has revealed in the Koran," and that Einstein "shared the thesis" that "Christian theology is a positive science."[83] Later, we're told quaint tales about Einstein keeping kosher and learning the Catholic catechism as a young boy—even though these have no bearing whatsoever on his later beliefs.[84] And a futile attempt to reconcile relativity with Christian theology fills the final hundred pages, even though Einstein always insisted that "my theory of relativity has nothing to do with theology."[85] Meanwhile, pantheism, pacifism, and Pythagoreanism—the triple keys, as we'll see, to Einstein's spirituality—don't even make it into the index. Maybe most egregious of all is Jammer's blatantly bogus claim that "Einstein never showed any interest in Far Eastern philosophy and never expressed any sympathy with Oriental religious thought or mysticism."[86] For all intents and purposes, then, there has yet to be any serious exploration of Einstein's spiritual side.

One of Einstein's favorite thinkers once said that "philosophy is a high mountain road," and we can thank Max Jammer for pointing the path out to us.[87] But like so many scholars before and since, he never really followed where it led. After ambling into the foothills of Einstein's philosophy, the dark clouds of an alien worldview began to descend on him, and he beat a hasty retreat back down to the lowlands of traditional theology.[88] He never so much as caught a glimpse of the mountain peaks of the mind where Einstein made his home.

If we wish to see the same vistas Einstein saw, we must climb higher. Einstein ascended often into what he called "the silence of high mountains, where the eye ranges freely through the still, pure air and fondly traces out the restful contours apparently built for eternity."[89] From these high places he plucked the finest wildflowers of world philosophy, assembling them into an artful arrangement all his own. A sensitive assessment of his spirituality must acknowledge all these manifold influences—ancient and modern, direct and indirect, East and West.

The way is long; we'd better get walking.

2

THE THIRD PHASE

There is a third phase of religious experience. I shall call it cosmic religious feeling.

—Albert Einstein, "Religion and Science" (1930)

E INSTEIN LONGED TO SEE THE HUMAN MIND SCALING THE HIGHEST mountains, reaching undreamt-of heights in science, ethics, and spirituality. And he was well aware that we'd already ascended quite a ways: there was much to admire in the arts and sciences, much to celebrate in our spiritual systems. But in surveying our situation, past and present, Einstein also saw that we tended to shirk the hard work. For thousands of years, we'd tried to fly on borrowed wings. Whether through miracles or revelations, prayer or the grace of God, humans had always secretly hoped to be *carried* up to Heaven. We'd sought out every imaginable shortcut to the sky, and the outcome was as obvious as it was odious: our minds were still stuck in the stubborn mud of superstition.

Einstein realized that there was only one way upward: the path from mud to mountaintop had to be *walked*, one small step at a time. Real wisdom would never be given; it had to be won. And we couldn't count on our received religions for much encouragement. Almost without exception, they agreed that the summit had been reached long ago by our spiritual superiors, who'd told us the ultimate Truth and given us God's final commandments. Making the ascent ourselves was not only unnecessary, but impossible: the path to the peak was too perilous for mere mortals.

But over the millennia, there'd always been an eccentric minority who were heralds of a more hopeful message. A select few, scattered across the centuries, had made the ascent and then come back down insisting that it wasn't enough to just marvel at the tales they told. If we wanted to see what they had seen, then we too must make the climb. Tradition tends to regard these teachers as heretics, but Einstein revered them as heroes. He knew that the genuine spirituality taught by "the true religious genius" was a joyous doctrine that "needed no dogma, no priestly caste, no humanized God."[1] There was nothing sinful in wanting to see with our own eyes, nothing sacrilegious in making the ascent ourselves.

The core of the cosmic religion was a commitment to keep climbing—not so we could look down on the religions of the past, but in order to gaze farther and maybe get a glimpse of the religion of the future. For most people, even the idea of spiritual progress is blasphemous. But Einstein was certain that the same inner strength that had inspired our ancestors was still alive in us. We had both the right and the responsibility to continue the ancient quest. There was no promise we'd reach the peak, but the destination was without doubt a worthy one. And the journey had only just begun.

Spirituality was not something static in Einstein's eyes. Religion was constantly evolving, always adapting to the new knowledge and novel needs of humanity. And in order to envision where spirituality might be going, Einstein thought it was essential to understand where it had come from in the first place. We had to appreciate the many metamorphoses the religious mindset had undergone over the millennia.

For the sake of simplicity, Einstein separated the human spiritual journey into three phases. The first phase was founded on fear. For our

Einstein's handwritten manuscript of his 1930 article "Religion and Science," written expressly for the *New York Times Magazine*. In this essay, he laid out his three-phase model of humanity's spiritual evolution and introduced the world to his third phase, the cosmic religion. The phrase "There is a third phase of religious experience; I shall call it cosmic religion" can be seen (in German, of course) in the second paragraph. The public reaction to Einstein's spirituality was largely hostile (see Chapter 5). Photo © the author (2023), used with permission of the Albert Einstein Archives.

early ancestors, Einstein thought that it was "above all fear that evokes religious notions—fear of hunger, wild beasts, sickness, death."[2] Early humans had no knowledge of the laws of nature, so it was easy for them to assume that when bad things happened, bad intentions had to be behind them. "Since at this stage of existence understanding of causal connections is usually poorly developed," Einstein explained, "the human mind creates illusory beings more or less analogous to itself on whose wills and actions these fearful happenings depend."[3]

And if malevolent forces were actively trying to undermine us at every turn, then the obvious way to prevent bad outcomes was by pleasing the gods. "Thus," Einstein continued, "one tries to secure the favor of these beings by carrying out actions and offering sacrifices which, according to the tradition handed down from generation to generation, propitiate them or make them well disposed toward a mortal."[4] Given the limited knowledge of the times, Einstein felt this was fair enough. After all, *something* had to be the cause of events, and so "the positing of causal animistic connections" was "not unreasonable in itself."[5] But even in ancient times, Einstein believed, there had occasionally been "minds who tried to interpret nature's law of cause and effect" without assuming the intervention of supernatural forces, proto-scientists who "gradually undermined the concept of the gods as angry and capricious beings."[6]

And so the human spirit came to a second phase, characterized by what Einstein called *social* or *moral* religions. "From the understanding of causal relations in creation evolved the concept of a God of justice," Einstein hypothesized. "He became humanized and comprehensible; man could look to Him for guidance and consolation."[7] This was "the God of Providence, who protects, disposes, rewards, and punishes," Einstein explained, "the comforter in sorrow and unsatisfied longing; he who preserves the souls of the dead."[8] For Einstein, all our familiar world religions fell into this second phase, and he admitted that it was undeniably an improvement.[9] "The development from a religion of fear to moral religion is a great step in peoples' lives," he thought.[10] "Nobody, certainly, will deny that the idea of the existence of an omnipotent, just, and omnibeneficent personal God is able to accord man solace, help, and guidance."[11]

Among many theologians and philosophers in Einstein's time, there was a complacent confidence that this was the climax of our religious evolution. Einstein disagreed. "For me," he wrote, "the Jewish religion, like all others, is the embodiment of the most childish superstition."[12] But besides being childish fantasies, second-phase religions had several other fatal shortcomings. For one thing, worldviews founded on what Einstein called "fairytales about creation" had a vested interest in fomenting blind faith and condemning open inquiry.[13] They couldn't be expected to honor the intellectual integrity so necessary to both the scientist and the true spiritual seeker. "Many concepts which generation after generation have considered absolutely true have limited or no validity," Einstein once mused. "That goes for the Church, too."[14] The man who single-handedly revolutionized physics knew the necessity of radical thinking, and knew firsthand that the faithful would always fight it.[15] "Look through the ages," he suggested, "at the array of scientists and philosophers who have been persecuted for their convictions."[16]

Just as bad as a narrow mind was a hypocritical heart. While Einstein credited second-phase religions for encouraging an elevated morality, he criticized them for failing to live up to their own lofty ideals. "Unselfishness, humaneness, service to your brother—these are the values which the Church should practice for once," Einstein complained, "instead of constantly trying to gather in more souls."[17] Worst of all was the wanton disregard for life itself. Einstein felt this was a failing among almost all the major faiths, but he himself was most familiar with Christianity's abysmal track record. "Consider the hate the Church manifested against the Jews and then against the Muslims," he once said, "the Crusades with their crimes, the burning stakes of the Inquisition, the tacit consent of Hitler's actions while the Jews...dug their own graves and were slaughtered."[18] For all their talk of love, when push came to shove religions looked out for their own interests, "supporting the mighty at the cost of the human rights of the poor."[19]

Einstein's overall assessment of our existing religions was dismal. Despite their redeeming qualities, they were hostile to science, unfriendly to free thought, and subservient to entrenched interests. So it's unsurprising that he saw transcending the second phase as not only desirable but necessary. What he envisioned in its place was a spirituality that went well beyond a paralyzing fear of capricious gods or a blind obedience to moral laws. And yet he had to admit that deeper spiritual development demanded a discipline that few people, and even fewer societies, could contend with. "Only individuals of exceptional endowments," he thought, "and exceptionally high-minded communities rise to any considerable extent above [the second] level."[20] And yet, rare though it might be, there *was* a third phase of religious experience to which we could and should aspire.

Einstein called it cosmic religion.[21] This third phase would transcend all the old artificial divisions: it would be a system that all people, in principle, could practice. And Einstein's most eloquent expression of its ideal was his aphoristic assertion, "I am a part of infinity."[22] This enigmatic feeling that every individual was an aspect of the Infinite was the essence of Einstein's religion of the future. And although it remains an audacious aspiration, it was just as much a revival of an age-old awareness. The phrase "third phase" might suggest a steady spiritual ascent since ancient times, but Einstein never argued for a stepwise progression from "primitive" to "higher" religion. On the contrary, the cosmic religious feeling was accessible at all times and in all places to receptive individuals.

How could an ancient feeling familiar to our ancestors also anticipate the philosophy of the future? An analogy with the scientific quest can help clarify this apparent conundrum. Einstein saw his science as revealing a small part—but a part nonetheless—of a great and eternal Truth, which knew no boundaries of language, culture, or custom. Obviously, his own scientific work would have been impossible without many recent innovations in physics and mathematics. But to Einstein, his real intellectual ancestry included all the great minds of the past who had struggled to understand our universe, those who had "shown the way to kindred spirits scattered wide through the world and the centuries."[23] He once confessed that "of all the communities available to us, there is not one I would want to devote myself to except for the society of the true searchers, which has very few living members at any one time."[24] For Einstein, all sincere seekers after scientific truth, few and far between though they might be, formed a fellowship that transcended time and space.

He saw genuine spirituality in exactly the same way. Far from being the exclusive property of any particular religion that had gotten it right and now had a monopoly on truth, Einstein believed that higher religious experiences could appear in any culture at any time—just like scientific truths. The cosmic religious feeling could flower even in the most unpromising soil. "The religious geniuses of all ages have been distinguished by this kind of religious feeling," Einstein wrote, "which knows no dogma and no God conceived in man's image; so that there can be no church whose central teachings are based on it."[25] Organized religions could rarely take credit for fostering such feelings. If anything, they tended to persecute the great spirits that arose in their midst. "Hence it is precisely among the heretics of every age," Einstein wrote, "that we find men who were filled with this highest kind of religious feeling and were in many cases regarded by their contemporaries as atheists."[26] And he once mused, with a twinkle in his eye: "I would have been in good company with those witches and heretics."[27]

Yet for all its historical precedents, the third phase wasn't simply a revival of religious sentiments promoted by the spiritual paragons of the past. A renaissance alone wasn't enough for Einstein. If religion was going to remain relevant into the future, it had to not only stir our most ancient spiritual impulses, but also integrate the new knowledge revealed by the sciences. Einstein was appalled at how second-phase religions tended to suppress science and persecute scientists in favor of their own fairy tales. "With his growing knowledge of the vastness of the universe and its trillions of stars," he thought, "man must consider it an insult when he is told that his conduct should be motivated by fear of punishment or hope of reward."[28] The cosmic religion, in contrast, would not only tolerate new knowledge, but actually *encourage* scientific inquiry. Einstein famously maintained that "the cosmic religious feeling is the strongest and noblest motive for scientific research."[29] What he meant was that the search for deep scientific truth was so demanding—and could cause so much friction with society (think of Galileo)—that only a kind of religious zeal could keep the quest in motion. And he offered his own little ode to the efforts made, and the ordeals suffered, by the pioneers of old:

> What a deep conviction of the rationality of the universe and what a yearning to understand...Kepler and Newton must have had to enable them to spend years of solitary labor in disentangling the principles of celestial mechanics!...Only one who has devoted his life to similar ends can have a vivid realization of what has inspired these men and given them the strength to remain true to their purpose in spite of countless failures. It is cosmic religious feeling that gives a man such strength.... in this materialistic age of ours the serious scientific workers are the only profoundly religious people.[30]

That final thought is a bold claim, but elsewhere Einstein made an even more brazen assertion: "True religion has been ennobled and made more profound by scientific knowledge."[31] Not only did he see a rarefied religious impulse as the main motivator of major scientific breakthroughs; he also suggested that science itself would, and should, come to serve as a source of religious wonder. "It seems to me," Einstein mused, "that science not only purifies the religious impulse of the dross of its anthropomorphism but also contributes to a religious spiritualization of our understanding of life."[32]

So discarding old dogmas didn't mean renouncing all religious sentiment. *Au contraire*: as we learned more and more about how marvelous reality really was, Einstein thought that a religious feeling was in fact the only reasonable response. "The further the spiritual evolution of mankind advances," he said, "the more certain it seems to me that the path to genuine religiosity does not lie through the fear of life, and the fear of death, and blind faith, but through striving after rational knowledge."[33] Filled with a deep reverence for the natural world and a hallowed respect for the human mind, the cosmic religion could be an encouraging companion on the scientific quest. And in turn, the ceaseless discoveries of science, witnessed with the right eyes, would provide a never-ending font of cosmic religious feelings.

Einstein's audacious notion of religion and science as reciprocal supports for a single spiritual quest is practically unprecedented.[34] But even if cosmic religion and scientific inquiry were destined to dovetail nicely in the near future, a Western scientific mindset was by no means a necessary precursor to the third phase. If anything, Einstein saw some of its first and finest examples in the Far East and the distant past. "The beginnings of cosmic religious feeling already appear at an early stage of development," he thought.[35] "Some of the Psalms and some Buddhist literature breathe this cosmic religion; so does the heathen Democritus, the Catholic St. Francis of Assisi, the Jew Spinoza."[36] It's a curious collection of names, among which Einstein also included Giordano Bruno and Mahatma Gandhi.[37]

These people could hardly have belonged to more divergent traditions; their countries and their cultures were separated by thousands of miles and sometimes thousands of years. It almost seems like Einstein simply picked some great names at random from the pages of religious history. But there's nothing arbitrary or accidental about his roster of religious geniuses. As Einstein saw it, the secret springs that fed the cosmic religious feeling were ancient and everywhere, and all these minds, despite the superficial differences, drank from the same sacred waters. All of them had transcended their times and their traditions and tapped into something more primordial.

With his third phase, Einstein wanted to return to the source of the supreme religious impulses that had sprung up spontaneously, far and wide, for thousands of years, and finally merge these many separate streams with the growing sea of scientific discovery. Yet compatibility with science was only one advantage of many. The most attractive aspect of the cosmic religious feeling was that it was accessible to anyone. First- and second-phase religions tended to devolve into desiccated institutions that inevitably created intermediaries between the individual and the divine. The old religions were "stabilized by the formation of a special priestly caste which sets itself up as a mediator between the people and the beings they fear," Einstein explained. "In many cases a leader or ruler or privileged class...combines priestly function with its secular authority in order to make the latter more secure; or the political rulers and the priestly caste make common cause in their own interests."[38] Einstein didn't mince words when it came to what should be done about this deplorable situation. "Superstition and priest rule are grave evils," he explained, that "must and should be fought."[39]

This battle had been ongoing for centuries, and by Einstein's time monotheism's monopoly on the European mind had started to crumble. Nietzsche's declaration that God is dead and Darwin's notion of natural selection were rapidly eroding the once-firm foundations of Western faith. "The influence of these mythic, authoritatively anchored forces," as Einstein put it, "had been reduced to a tolerable level in spite of all the persisting inertia and hypocrisy."[40] For perhaps the first time in history, atheism was becoming acceptable, or at least tolerable (no longer a reason to be burned alive, even if it was often still enough to be ostracized).

At the same time, exposure to foreign forms of spirituality was becoming more and more common. It was inevitable that Einstein would encounter traditions in which every individual had both the power and the prerogative to pursue their own enlightenment, more or less free from the meddling of pesky priests. And to the man who was always such a staunch advocate of individual liberty, especially freedom from religious and political propaganda, these traditions would prove irresistible.[41]

Einstein existed at a unique moment in European history. For the first time, Eastern spiritual texts, and even the spiritual teachers themselves, were now widely available in the Western world. Up until the 1600s, Europe and Asia had long been separated by both linguistic and geographical barriers. Cultural exchange was mediated mostly through the slow-motion, low-fidelity conduit that was the Silk Road.[42] But beginning in the seventeenth century, long-distance seafaring brought the two continents into continuous contact. Christian missionaries went east en masse to "convert the heathens," but thoughtful theologians and sensitive scholars soon realized that they were confronting a surprisingly sophisticated (maybe even superior) spirituality.[43] The first trickles of ancient Eastern wisdom began flowing westward, and then the trickle turned into a torrent. It wasn't long before many major works were available in translation. The Upanishads and the Tao Te Ching, Buddhist sutras and the Bhagavad Gita began to bring about a slow but steady metamorphosis in the Western mind.[44]

One of the first to become thoroughly acquainted with these new forms of thought was Einstein's second-favorite philosopher, Arthur Schopenhauer.[45] From the very first page of his 1818 magnum opus, *The World as Will and Representation*, he was already praising ancient Indian thinkers for foreshadowing his own philosophy.[46] Actually, even prior to the first page, in his preface, he went as far as arguing that access to the Upanishads and the Vedas was "the greatest advantage which this still young century has to show over previous centuries."[47] In other words, Schopenhauer was so awed by ancient Indian thought that he saw its entry into Europe as quite literally the event of the century.

Schopenhauer lived what he loved. He slept with Hindu scriptures by his bedside, owned an ancient statue of the Buddha, and even named his dog Atman (a Sanskrit word for the inner aspect of every individual that is identical to the divine).[48] His masterwork, *The World as Will and Representation*, made hundreds of references to Oriental scriptures. The direct quotations are so copious and his commentary so astute that his book actually doubles as a decent introduction to Eastern philosophy as a whole.

Einstein called Schopenhauer's writings "wonderful," and he so admired the man that he had a portrait of him hung in his Berlin study.[49] He owned a twelve-volume set of the philosopher's complete works, which he read over and over again—and by all accounts with great pleasure.[50] So when Einstein spoke of Buddhism as a prototype for the spirituality of the future, he was only echoing Schopenhauer, who thought that Eastern wisdom would revolutionize European thinking in the same way that the rediscovery of classical culture had during the Renaissance.[51] "I surmise," he wrote in 1818, "that the influence of Sanskrit literature will penetrate

Einstein's edition of the complete works of the German philosopher Arthur Schopenhauer, at the Albert Einstein Archives in Jerusalem. Schopenhauer was one of the first Western philosophers to thoroughly acquaint himself with the newly translated texts of Eastern spirituality. He venerated the Vedas as "the fruit of the highest human knowledge and wisdom," and in 1818 he argued that access to the Upanishads and the Vedas was "the greatest advantage which this still young century has to show over previous centuries." He filled his books with hundreds of direct quotations from Eastern scriptures, focusing especially on the nondualistic thinking of the Upanishads. For Schopenhauer, the ultimate aim of philosophy was neither rational understanding nor logical consistency, but the attainment of "that peace that is higher than all reason, that ocean-like calmness of spirit, that deep tranquility, that unshakable confidence and serenity." And he made it clear that he considered this kind of enlightenment as analogous to "reabsorption in Brahman, or the Nirvana of the Buddhists." Like many other scientists of his generation, Einstein's first exposure to Eastern spirituality was through Schopenhauer. Einstein called Schopenhauer's writings "wonderful" and read his collected works over and over again. He so admired the man that he had a portrait of him hung in his Berlin study. Photo © the author (2023), used with permission of the Albert Einstein Archives.

no less deeply than did the revival of Greek literature in the fifteenth century."[52]

Schopenhauer's prediction seems awfully prescient when we consider the enormous effect Eastern thought and spiritual practice are now having on almost every aspect of the modern Western world.[53] And yet this contemporary influence is only the latest incarnation of an accelerating but by no means novel trend. When the quantum physics revolution came along at the turn of the twentieth century, the influx from the East had been well under way for some time. So it's no mystery how Eastern philosophy ended up exerting a profound influence not just on Einstein, but also on the other founders of the new physics.

Almost all the architects of quantum mechanics yearned for some new synthesis of science and spirituality; many felt something resembling Einstein's cosmic religious feeling; and most became intrigued by the traditions of the East, which seemed to anticipate their own scientific insights and spiritual intuitions. Werner Heisenberg once told an astonished interviewer that "his own scientific work had been influenced, at least at the subconscious level, by Indian philosophy."[54] Wolfgang Pauli knew that "Western thought as a whole has always been influenced by the near and far Asiatic East," and summed up his spiritual pedigree in simple terms: "When it comes to religion and philosophy, my background is Lao-tse and Schopenhauer."[55] Niels Bohr thought that physicists were now facing "those kinds of epistemological problems with which already thinkers like the Buddha and Lao Tzu have been confronted," and the family coat of arms he designed bore the Taoist yin-yang symbol.[56] Robert Oppenheimer had a lifelong interest in Eastern thought and learned Sanskrit in his spare time; he was adept enough to read the Bhagavad Gita in the original and to try his hand at translating it himself.[57] And Erwin Schrödinger read countless tomes on Eastern religion and savored every word Schopenhauer wrote.[58] He even seems to have experienced a kind of spiritual awakening showing him that he was a part of Infinity. "[You] suddenly come to see, in a flash, the profound rightness of the basic conviction in Vedanta," he wrote. "That *you* are a part, a piece, of an eternal, infinite being, an aspect or modification of it...Hence this life of yours which you are living is not merely a piece of the entire existence, but is in a certain sense the *whole*."[59]

Obviously, Einstein's admiration for the East was anything but extraordinary for a physicist of the times. Be it Hinduism or Taoism, Buddhism or the Bhagavad Gita, the founders of quantum theory were unanimous in their esteem for Eastern wisdom. And these men were not merely Einstein's fellow physicists. They were friends and (intellectual) foes, lifelong colleagues and sometimes direct collaborators.[60] If Einstein had actually had no interest in Eastern spirituality, as one scholar insisted, he would have been the inexplicable exception to a proclivity that was pervasive among his peers.[61]

Many of the founders of the new physics even embarked on lengthy voyages to the Orient at a time when intercontinental travel was hardly for the faint of heart.[62] Einstein's own journey to the East is little known today but well documented in his travel diaries. When he heard he'd won the Nobel Prize in 1922, he didn't bother to accept the award in person. He'd already been invited to visit Japan and had "immediately resolved to embark on such a great voyage that must demand months."[63] It was an easy decision. "I would never have been able to forgive myself," he felt, "for letting a chance to see Japan with my own eyes pass unheeded."[64] So he skipped the Nobel ceremony in Sweden and instead embarked on a six-month trip to Asia to satisfy his "yearning for the Far East."[65]

His unfiltered travel diaries reveal that the dinners with dignitaries and seminars with scientists left him drained and desperate to escape to his hotel room. He had no patience for the "endless handshaking" and "schmaltzy speeches," the "daft questions" and "interminable" banquets.[66] But everywhere he went, he displayed a deep interest in Eastern religion. He took the time to visit Buddhist temples in Sri Lanka, China, and Malaysia, as well as dozens of "splendid" and "magnificent" sites in Japan—including even some Shinto shrines.[67] The trip must have made an impression, because Einstein would later pack his personal library with books on Eastern spirituality.

While most of the Western world remained scornful of Eastern religion or ignorant of it altogether, Einstein was traveling all over Asia, hanging around Buddhist monasteries, chatting with monks about meditation, and reading Taoist and Buddhist scriptures in his spare time. And yet for decades the ample evidence that Einstein was inspired by Eastern spirituality has been downplayed, denied, and sometimes simply deleted.[68] When Einstein tells us that he "agree[s] with the aims of vegetarianism for aesthetic and moral reasons" and has "long been an adherent to the cause in principle," experts insist that "Einstein was probably not a vegetarian by choice, for he left behind no remarks that it was a moral issue for him."[69] When Einstein urges us to "free ourselves from this prison [the ego] by widening our circle of compassion to embrace all living creatures," we're told that he's speaking "not [of] the passive compassion found in Eastern religion...but the active compassion of the Biblical prophet."[70] And when Einstein says point-blank that "Buddhism contains...a much stronger element" of the cosmic religious feeling than Judaism, and points to the Buddha as "a person who is religiously enlightened," the foremost scholar of Einstein's spirituality can still somehow claim that "Einstein never showed any interest in Far Eastern philosophy and never expressed any sympathy with Oriental religious thought or mysticism."[71]

To borrow a quip from Einstein's friend and fellow physicist Wolfgang Pauli: these flagrant falsehoods are "not even wrong." Einstein was always open about his admiration for Eastern spirituality. There was no person he admired more ardently than Mahatma Gandhi;[72] no virtue he valued above the Indian ideal of nonviolence, *ahimsa*;[73] and no enlightenment he held in higher esteem than experiencing firsthand the ancient axiom of the Upanishads, *Tat tvam asi*: consciousness and cosmos are one.[74] But even though Einstein's esteem for Eastern spirituality was obviously genuine, that doesn't mean he was an expert. He didn't learn any Asian languages, he didn't study Oriental scriptures in any detail, and he didn't engage in any of the meditation practices so central to most Eastern spiritual paths.

Nor was he in any sense a convert. Rather, his cosmic religion shared a sort of family resemblance with Eastern spiritual systems, which provided profound inspiration, but by no means a perfect model, for his third phase. Generally speaking, Eastern traditions agreed with Einstein's emphasis on the nonduality of Nature, on the need for an all-inclusive ethics embracing every living thing, and on the importance of an expansion of consciousness in the individual's quest for some kind of enlightenment. And unlike Western religions, with their blind faith, their anthropomorphic God, and their historical hatred of science, Einstein found far less in Eastern religions to offend his independent mind.

But even the Eastern religions Einstein most admired came with considerable cultural and metaphysical baggage, and these were not burdens Einstein intended to keep carrying. The third phase was a summit we still hadn't scaled, a destination that could be reached only through an austere discipline of the spirit. To reach this peak, we had to travel light. And so Einstein insisted on keeping his spirituality simple. When a Christian minister once asked him bluntly what he meant by cosmic religion, he offered a concise synopsis of the new creed. "It is not a religion that teaches that man is made in the image of God—that is anthropomorphism," Einstein replied. "This religion has no dogma other than teaching man that the universe is rational and that his highest destiny is to ponder it and co-create with its laws."[75]

Was this really even religion? Most people in Einstein's era didn't think so, and many people still don't today. In Einstein's time, there were widespread accusations that the cosmic religion was no different from atheism—a misconception that extended from the average American believer all the way up to the highest echelons of religious orthodoxy.[76] And even now, prominent unbelievers like Richard Dawkins argue that Einstein's pantheism is nothing more than "sexed-up atheism."[77]

It's easy to understand these sentiments. With its disdain for traditional dogma, its denial of a personal God, and its insistence on using reason rather than revelation to understand reality, Einstein's third phase looks an awful lot like atheism at first glance. And actually, at *second* glance it kind of does, too. A close reading of Einstein's scattered writings on spirituality can convince us that, for him, not only was there no personal

God—there was also no prayer,[78] no sin,[79] and no Savior;[80] no Heaven or Hell;[81] no divine rewards or punishments;[82] and no higher sanction at all for human morality.[83] It's hard to disagree with Richard Dawkins that "Einstein was, in every realistic sense of the word, an atheist."[84]

But if Einstein was an atheist in letter, he certainly wasn't in spirit. Atheism might be an adequate summary of many people's worldview— many scientists, in particular—but it's a purely negative position. It tells us only what someone *doesn't* believe. And as Einstein himself asserted, "Mere unbelief in a personal God is no philosophy at all."[85] What Einstein *did* believe still remains to be discovered, but whatever it was, he didn't think the atheist label did it justice ("I am not an atheist" is about as clear a statement on the matter as one could hope for).[86] Einstein also took pains to distance himself from atheism in other public pronouncements. "In view of such harmony in the cosmos which I, with my limited human mind, am able to recognize, there are yet people who say there is no God," he once said. "But what makes me really angry is that they quote me for support of such views."[87] It might seem like that settles the matter.

Alas, it's not so simple. For one thing, Einstein did call himself an atheist at times, or at least recognized that others might reasonably do so. In 1945, a young US Navy ensign wrote to Einstein asking if it was true that the great physicist had met a Jesuit priest and been converted to Catholicism. "I have never talked to a Jesuit priest in my life," an incredulous Einstein replied, "and I am astonished by the audacity to tell such lies about me. From the viewpoint of a Jesuit priest I am, of course, and have always been an atheist."[88] Although he didn't really like the label himself, Einstein readily acknowledged that, from the standpoint of a standard believer, an atheist is exactly what he was.

We also know that Einstein deliberately went against the tradition of the times—which expected everyone to identify with some religious denomination—even when the stakes were high. When he applied for his first real professorship in 1910, he refused to identify as Jewish; he boldly listed his religious affiliation as "None."[89] When he was granted a divorce from his first wife, in 1918, he entered his religion as "dissenter" in his deposition to the court.[90] And when he moved to Berlin in 1920 to take up a prestigious post as the world's most famous physicist, he declined to join the city's official Jewish community. "Nobody can be compelled to join a religious community," he reminded them. "Those times, thank God, are gone forever.... [I] shall remain unassociated with any official religious group."[91] Given the gravity of all three situations, it's impossible to imagine he was just playing the joker. He must have felt he was making a meaningful statement.

And if Einstein was actually appalled by atheism as such, then he kept awfully strange intellectual company. Almost all his favorite thinkers and philosophers were atheists: Democritus, Spinoza, Hume, Goethe, Schopenhauer, Nietzsche, and Ernst Mach. Einstein even described atheists like Democritus and heretics like Spinoza as "religious geniuses."[92]

This fondness for infidels wasn't confined to the thinkers of the past; it also embraced famous atheists in the present. Einstein was an especially great admirer of Bertrand Russell: master mathematician, prolific polymath, winner of the Nobel Peace Prize—and outspoken atheist.[93] In 1953, Russell published an essay, as controversial as it was concise, entitled "What Is an Agnostic?" "An Agnostic may hold that the existence of God, though not impossible, is very improbable," Russell explained to his readers. "He may even hold it so improbable that it is not worth considering in practice. In that case, he is not far removed from atheism." He went on to make a cheeky comparison between the biblical God and the Olympian gods of ancient Greece. "If I were asked to *prove* that Zeus and Poseidon...do not exist, I should be at a loss to find conclusive arguments," he admitted. "An Agnostic may think the Christian God [just] as improbable."[94]

Einstein loved the essay. "Bertrand Russell's article on religion," he wrote to a friend, "is masterfully crafted—everything makes sense."[95] And Einstein didn't keep his praise private. When Russell was offered a faculty position at the City University of New York, conservative Christian groups fought back and brought a legal suit against him. The courts pronounced Russell "morally unfit" to teach, and the job offer was withdrawn. Einstein was incensed and penned an open letter in Russell's defense. It was published in the *New York Times* and included what would become one of his most famous turns of phrase. "Great spirits have always encountered violent opposition from mediocre minds," he wrote. "The mediocre mind is incapable of understanding the man who refuses to bow blindly to conventional prejudices and chooses instead to express his opinions courageously and honestly."[96] High praise indeed from a man who allegedly abhorred atheists.

Still, there was something about atheism that Einstein found unsettling. In 1952, a woman who'd escaped from a life of rigid religious orthodoxy wrote Einstein a letter begging him to clarify his views. "It means a great deal to me to know," she explained, whether Einstein, too, was "a free thinker, possibly atheist or agnostic."[97] Part of Einstein's reply is well known: "The idea of a personal God is quite alien to me and seems even naïve."[98] But the rest is even more revealing:

> However, I am also not a "Freethinker" in the usual sense of the word because I find that this is in the main an attitude nourished exclusively by an opposition against naïve superstition. My feeling is religious insofar as I am imbued with the consciousness of the insufficiency of the human mind to understand deeply the harmony of the Universe which we try to formulate as "laws of nature." It is this consciousness and humility I miss in the Freethinker mentality.[99]

So it wasn't atheism per se that Einstein disapproved of. It was "the fanatical atheists whose intolerance is of the same kind as the intolerance of the religious fanatics and comes from the same source." Einstein called such people "creatures who—in their grudge against the traditional 'opium for the people'—cannot hear the music of the spheres."[100]

For Einstein, these fanatical atheists were oblivious to the awesome mystery before which we found ourselves, blind to what he called "the Wonder of nature."[101] Although Einstein had little respect for received religion, he saw fanatical atheism as a classic case of the cure being worse than (or, anyway, as bad as) the disease. "I myself would never be part of such an undertaking," he said, "because such a belief [in a personal

God] seems to me still better than the lack of any kind of transcendental interpretation of life. And it seems doubtful whether one can offer to most men with any success a sublime means to satisfy the metaphysical need."[102] Einstein knew people's metaphysical needs had to be met, and since "sublime means" were too much to ask of most, pragmatic methods would have to do for the time being.

"By virtue of its simplicity," Einstein realized, the idea of a personal God was "accessible to the most undeveloped mind," and accessibility had its advantages. "But on the other hand," he continued, "there are decisive weaknesses attached to this idea in itself, which have been painfully felt since the beginning of history."[103] Einstein's acceptance of traditional faith was not a matter of tolerance, then, or even agnosticism, but rather resignation. In a letter written in 1920, he lamented that "even nowadays, eliminating the sacred traditions would still mean spiritual and moral impoverishment—as gross and ugly as the attitude and actions of the clergy may be in many respects."[104] And so, although he rejected monotheism in principle, he accepted people's faith in a personal God as a necessary evil (or expedient) in practice. Some scholars, such as Max Jammer and the theologian Alister McGrath, have interpreted this stance as tacit support for traditional religion, but "better than nothing" is pretty faint praise.[105]

Ultimately, what worried Einstein wasn't unbelief in God, but the absence of any big-picture perspective at all. He abhorred *nihilism*, not atheism. For him, a life lived without a sense of wonder and purpose was no life at all. "What is the meaning of human life, or for that matter, of the life of any creature?" he once asked. "To know an answer to this question means to be religious.... The man who regards his own life and that of his fellow creatures as meaningless is not merely unhappy but hardly fit for life."[106]

As a culture, we might feel that we've adequately assimilated the hard lessons of physics and philosophy over the last few hundred years: God is dead, Nature is probabilistic, nothing is true, everything is permitted. And perhaps we have become sufficiently skeptical, secular, and cynical. But a naïve, rather nasty nihilism was not the endgame Einstein had in mind. His third-phase spirituality was more than just "sexed-up atheism" or "watered-down theism."[107] Einstein saw that a genuine sense of awe was in short supply among complacent believers and fanatical atheists alike. And this ephemeral feeling was so important to him that he would make wonder the central axis around which his entire spirituality revolved.

3

WORLD OF WONDER

The most beautiful thing we can experience is the mysterious.... He to whom this emotion is a stranger, who can no longer pause to wonder and stand rapt in awe, is as good as dead.

—**Albert Einstein, "What I Believe" (1930)**

THE COSMIC RELIGION WAS WOVEN FROM A FABRIC OF PURE WONDER—and anyone who was in awe of our world was welcome. Einstein dispensed with all the standard demands: no specific beliefs were required, no sacred book had an exalted status, and no particular race of people got special treatment. He severed his third phase from the religious paradigms of the past by instead making wonder the starting point and central focus of his spirituality.

Not that Einstein's emphasis on emotion was anything new. He recognized that "the most varying emotions preside over the birth of religious thought and experience," whether those emotions were positive, negative, or neutral.[1] But every previous religion presumed that we knew our weird world well enough to know the one right way of acting within it. Some insisted life's challenges must be met with love;[2] some saw life itself as a terrifying test of faith;[3] and some concluded our world wasn't worth it at all—that both the sour and the sweet should be transcended through tranquil contemplation.[4] All these approaches, diverse as they appear, agree that the ultimate answers were found thousands of years ago. They give us final judgments and final goals—petrified moral codes for petrified, mortal souls.

But for Einstein, "to wonder and stand rapt in awe" was "the center of true religiousness."[5] Rather than seeing the world as fundamentally good or bad, he saw it as first and foremost *mysterious*. Wonder might seem like a weird foundation for a spiritual system, but it was just the kind of wide-eyed emotion Einstein wanted for his open-minded religion. Wonder is all about curiosity and questioning and questing. It reveals that existence is astonishing, and reminds us of the great enigma when we become complacent. It insists on our ignorance, provokes honest inquiry, and sustains our searching. Wonder represents a beginning, rather than an end; a call to adventure, rather than a conclusion. And for Einstein, real religion was impossible without it.

In emphasizing awe, Einstein was parting ways with most past religious teachers, but he still had plenty of predecessors. Socrates said some twenty-five hundred years ago that "wonder is the mark of the philosopher."[6] Schopenhauer saw "the sense of the sublime" as a sure sign of a higher mind.[7] And one of Lao Tzu's last lessons in the Tao Te Ching is "Let not your consciousness of life become shallow, and never allow yourself to become weary of existence."[8]

Aligning himself with all these first-rate philosophers, Einstein maintained that mere existence was marvelous. "Every thinking person," he felt, "must be filled with wonder and awe just by looking up at the stars."[9] And as it turned out, even our familiar starry firmament was more amazing than anyone had ever imagined. Einstein's own physics revealed that we lived in a weird wonderland where time could be dilated, light could be bent, and the endless sea of stars above had burst forth from a single mysterious seed in the Big Bang billions of years ago.

But wonder wasn't only accessible above or beyond: it was in front and below and all around us, erupting out of every little nook and cranny of our cosmos. "If we look at the tree outside whose roots search beneath the pavement for water, or a flower which sends its sweet smell to the pollinating bees, or even our own selves and the inner forces that drive us to act, we can see that we all dance to a mysterious tune," Einstein once said. We could call it whatever we wanted—"Creative Force, or God"— but the only certainty was that the wondrous force woven through all things everywhere "escapes all book knowledge."[10] For most of human history, wonder had been reserved for the kings and emperors who ruled the world, or for supernatural forces believed to control it from beyond. Einstein never asked us to abandon our awe, but he did want us to relinquish our respect for races and religions, nations and gods. He wanted a reorientation of reverence toward the world right in front of us: particles and people, plants and planets, the whole colorful ribbon of reality.

And maybe even more marvelous than the world all around us was the world within. For Einstein, the most astonishing thing of all was the power of our own minds. Much like existence itself, most people took the mind for granted. But Einstein discerned a deep mystery in the efficacy of our mental models. "The most incomprehensible thing about the universe," he insisted, "is that it is comprehensible."[11] Somehow we fashioned a functional rendition of reality that helped us "orient ourselves in the labyrinth of sense impressions." But in the final analysis, it was all "a free creation of the human mind," and we were "never completely guaranteed" that our experience of existence was anything more than "an illusion or hallucination."[12] To the minds of ancient India, this was the magic show of maya. Immanuel Kant called it "mere phenomenon" and famously argued that we had to resign ourselves to being forever ensnared by our mental representations. But Einstein disagreed, and what made him marvel was precisely our ability to see past our cognitive constructs and peer into the core of existence. "A priori," he said, "one should expect a chaotic world which cannot be grasped by the mind in any way."[13] But his own experience proved otherwise: the mind's powers were prodigious.

Einstein saw the first faint glimmer of this power as a young boy, when he started studying geometry. "By the time I was twelve," he tells us, "a thin schoolbook on geometry had become my most holy possession."[14] Within this "little book dealing with Euclidean plane geometry," he was astonished to find assertions that, "though by no means evident, could nevertheless be proved with such certainty that any doubt appeared to be out of the question. This lucidity and certainty made an indescribable impression upon me."[15] The mind's extraordinary efficacy was an epiphany that led Einstein to "wonder of a totally different nature."[16] He thought it was "marvelous" that we were "capable at all of reaching such a degree of certainty and purity in pure thinking as the Greeks showed us for the first time to be possible."[17] Euclid had shown him just "how powerful an instrument the human mind could be."[18] Before long, he started to see that the mind might even be able to move beyond the veil of maya entirely.

It wasn't simply idle curiosity that sustained this search. Einstein maintained that a "mystical drive" was what made him yearn to discern the essence of existence, and he openly acknowledged that his "rapturous amazement" for our awesome world was "beyond question closely akin to that which has possessed the religious geniuses of all ages."[19] But he rejected revelation and had nothing but scorn for the supernatural. "My religion," he said, "is to use my thinking faculties, as much as I can, to know what seems unknowable."[20] Venturing beyond the veil of maya meant pushing the mind past its ordinary limits. There could be no divine grace, no gifts from the gods. Oneness had to be won the hard way.

The reward was what Einstein called "cosmic religious feeling." It was all well and good to have an abstract intuition that all things were One, but wonder went further. It conveyed the unmediated *feeling* of being a part of Infinity, providing a powerful emotional counterpart to our limited intellectual comprehension. "I like to experience the universe as one harmonious whole," Einstein once said. "I have cosmic religious feelings.

I never could grasp how one could satisfy these feelings by praying to limited objects. The tree outside is life, a statue is dead. The whole of nature is life."[21] In opening our eyes to the omnipresent light of Nature, wonder woke us up, widened our minds, and conferred a new worthiness on our otherwise limited lives. And since we had no other way of knowing this deeper reality directly and definitively, wonder was our most precious possession. "He to whom this emotion is a stranger," Einstein wrote, "who can no longer pause to wonder and stand rapt in awe, is as good as dead: his eyes are closed."[22]

Einstein wanted our eyes wide open. His ardent desire was to share the wonder that had served him so well and have us all "stand like curious children before the great Mystery into which we were born."[23] And a childhood wonder was in fact where his own awakening had begun. "A wonder of this kind I experienced as a child of four or five years, when my father showed me a compass," he recalled late in life.[24] "The movement of the needle so intrigued me that I could hardly sleep. I wanted to find out why the needle never deviated, even when I turned the compass round and round."[25] Right up until the end of his life, he always insisted that "this experience made a deep and lasting impression upon me. Something deeply hidden had to be behind things."[26]

But what? Science had no satisfying answers. Einstein's father admitted that no one knew what this hidden force really was, and his uncle Jakob agreed that magnetism was a mystery. Yet Uncle Jakob encouraged Einstein's questioning. "What you don't know, call x," he told his nephew, "then hunt until you find what it is." The advice would prove decisive. "From that time on," Einstein explained, "I have called everything I didn't know x."[27] The Earth's mysterious magnetic field had determined more than just the direction of a compass needle; the destiny of a curious young mind had also been decided. Einstein's hunt for the hidden harmony of reality was on.

A few years later, he read a series of popular science books with what he called "breathless attention."[28] One volume after another opened his mind to the many other marvels of the natural world.[29] "The more I read," Einstein said, "the more puzzled I was by the order of the universe and the disorder of the human mind."[30] Here Einstein encountered the figure of the scientist, whose life was spent wrestling with the awesome mysteries of Nature. And he realized his awe needn't be just an occasional experience; wonder could be a way of life. "The consequence was a positively fanatic orgy of freethinking," he later recalled. "Out yonder there was this huge world...which stands before us like a great, eternal riddle....The contemplation of this world beckoned as a liberation."[31]

He didn't take his liberation lightly. Solving the riddle of the world was the "supreme goal," and he knew that deep questioning came with consequences.[32] Standing in awe of Nature might provide plenty of profound feelings, but it offered none of the easy answers of established religion. Unquestioning believers could find quick comfort in ready-made responses to almost any conceivable question; the serious seeker, on the other hand, had to take greater risks to secure the uncertain reward of a personal revelation. Einstein surveyed the great minds who had walked this path before him and decided that, for all its downsides, it was worth pursuing. "Similarly motivated men of the present and of the past," he figured, "were the friends who could not be lost."[33] And by the end of his life, he was confident the wager had been worthwhile. "The road to this paradise was not as comfortable and alluring as the road to the religious paradise," he wrote, "but it has shown itself reliable, and I have never regretted having chosen it."[34]

For Einstein, the "sense of the mysterious" was "the most beautiful thing we can experience," but it wasn't just all the fine feelings along the way that made the journey worthwhile.[35] Much more important was that wonder kept the mind in continuous motion. It was the ultimate wellspring of creative work, "the source of all true art and science"[36]—at least for those lucky enough to be born with a knack for numinous feelings. For a mind like Einstein's, the constancy of a compass needle was enough to awaken a lifelong sense of awe. But he knew we weren't all so well attuned to the natural world's wonders. "How can cosmic religious feeling be communicated from one person to another," he worried, "if it can give rise to no definite notion of a God and no theology?" For those who were numb to Nature's charms, Einstein thought, the achievements of other human minds should still be sufficient to shake us awake. "In my view," he wrote, "it is the most important function of art and science to awaken this feeling and keep it alive in those who are receptive to it."[37] Creative works were like the wonder once felt by other minds made manifest and materialized—a way for slumbering spirits to see for a while through awakened eyes.

And never in history has any mind inspired as much awe as Einstein's. His international fame first came in 1919, after Arthur Eddington's eclipse observations confirmed that the great gravitational effects of our sun bent beams of light coming to us from distant stars. But despite this very visible evidence that Einstein's theory was right, rumor had it that general relativity was so complex that only three people in the world understood it (two of them being Eddington and Einstein). When someone asked Eddington if the rumor was true, and a pained expression came over his face, they reassured him that "there's no need to be embarrassed, Professor, you are much too modest." Eddington protested: "It's not a question of modesty. I was only asking myself who the third could be."[38]

And yet utter incomprehension didn't seem to diminish the average person's awe at all. Somehow people sensed that Einstein's mind was a kind of miracle, even if they couldn't understand its creations. He became one of the wonders of the world, the first modern celebrity scientist—and he remains the most revered and most recognizable physicist of all time. Never before and never since has anyone been so renowned by so many for such esoteric achievements. The hubbub bordered on mass hysteria. "Since the light deflection result became public," Einstein wrote to a friend, "such a cult has been made out of me that I feel like a pagan idol."[39] And it wasn't just average people who were amazed by his mind. Even jaded intellectuals were in awe of his genius. The novelist C. P. Snow held that "he was in many ways different from the rest of the species."[40] A historian of science once called him "one of the wonders of Nature."[41] And one journalist who sat down to chat with him said that "it was like going to tea with god."[42]

Those in the best position to judge his gifts—physicists, mathematicians, Nobel laureates—agreed. Einstein wasn't just *a* genius; he was the one who inspired wonder even among the most marvelous minds of his time, "outstanding even among the most eminent," as Bertrand Russell put it.[43] Max Planck predicted that "in the centuries to come Einstein will be celebrated as one of the brightest stars that ever shined."[44] Louis de Broglie (another Nobel laureate) contended that "one cannot contemplate without astonishment and admiration work at once so profound and so powerfully original."[45] And Einstein's friend Abraham Pais assures us that even "[Wolfgang] Pauli, not known for an excess of awe, was...different in Einstein's company. One could perceive his sense of reverence. [Niels] Bohr, too, was affected in a similar way."[46] Arthur Eddington's assessment is no exaggeration: "Einstein stands above his contemporaries even as Newton did."[47]

It wasn't just his intellectual gifts that garnered Einstein this godly reputation. According to J. Robert Oppenheimer, "there was always with him a wonderful purity."[48] His down-to-earth simplicity, his defense of the oppressed, his desire for peace, and his disdain for wealth and honors all contributed to the otherworldly aura about him. Not long before he died, a birthday note from Princeton's professors revealed some other reasons for the reverence. "You have long dwelt in our midst," it read, "as a very near and personal reminder of our highest aspirations."[49] For many, Einstein exemplified not only what a human mind might achieve, but what a human being might become. Although his cosmic religion failed to find adherents and fill their minds with wonder, the man himself succeeded in spectacular fashion.

And yet Einstein was acutely uncomfortable with his own apotheosis. "It is an irony of fate that I myself have been showered with so much uncalled-for and unmerited admiration and esteem," he wrote.[50] "It strikes me as unfair, and even in bad taste, to select a few individuals for boundless admiration, attributing superhuman powers of mind and character to them."[51] But something more than just humility was haunting him. Although Einstein thought it an excellent thing for exceptional individuals to inspire others, he also understood there was a fine and fateful line between inspiration and idolization.[52] "The world needs heroes," Einstein acknowledged, "and it's better they be harmless men like me than villains like Hitler."[53] But better still would be for us to stop deifying individuals entirely.

And so Einstein insisted that there was "no room" in real religion "for the divinization of a nation, of a class, let alone of an individual."[54] He was well aware that when wonder was lavished on what was least worthy—a master race, an expansionist state, a despotic dictator—the consequences were as predictable as they were appalling. Nowhere was this more obvious in his time than in the unbounded admiration for Adolf Hitler. In July 1939, just weeks before war broke out, Einstein wrote to a friend that the German people had "always had the tendency to serve psychopaths slavishly." Horrified by what he saw on the horizon, however, he added that "they have never been able to accomplish it so successfully as at the present time."[55]

It seemed that some of our most sublime emotional impulses were highly susceptible to the opportunism of the unscrupulous—and Einstein wasn't the only one perplexed by this paradox. The atrocities of the Second World War set off a surge of research by psychologists seeking to understand why wonder and awe were so easily abused by authority figures. Scientists have concluded that awe is an ancient and sometimes awful emotion. A weird fusion of fear and wonder, it's thought to have arisen early in primate evolution as an emotion that promoted submission in the presence of more powerful peers. A subordinate's awed respect for a superior—such as a dominant alpha male—could have helped cement the social hierarchies so central to primate societies.[56] And if awe really did originate in primate dominance hierarchies, then feelings of wonder would forever be susceptible to exploitation.

With his worship of wonder, Einstein knew he was playing with powerful fire. But he also saw a simple way of avoiding abuses of awe: wonder had to be coupled with curiosity. While wonder alone would always be vulnerable to exploitation, genuine curiosity required both intellectual integrity and individual liberty. Neither quality was compatible with being a docile accomplice of religious or political despots. Or as Einstein put it, "A strong critical spirit prevents blind obeisance to any mortal authority."[57]

Although awe appears to have somewhat sordid animal origins, Einstein saw our inborn inquisitiveness as much more than another primitive instinct. For Einstein, this compulsion to comprehend our cosmos was quasi-divine, "a mystical drive in man to learn about his own existence."[58] He maintained that "the miracle of the human mind" was that it could "use its constructions, concepts, and formulas as tools to explain what man sees, feels, and touches."[59] And if the mind was a kind of miraculous instrument for interpreting reality, then curiosity was the main motive force pushing us to put its powers to the highest possible use.

For the cosmic religion, curiosity was therefore the sine qua non of the sincere spiritual seeker. And like his worship of wonder, Einstein's idolization of inquisitiveness set his spirituality apart from traditional systems. First- and second-phase religions had always found freethinking individuals intractable—and hence unwelcome. Moses the Lawgiver had no need for the lawless. Jesus the Good Shepherd wanted sheep, not wolves. Mohammed the Messenger brought a message of submission.[60] And since confident, creative individuals inevitably threaten entrenched ideology, concerted efforts to crush curiosity have always been a time-honored religious tradition.

Nowhere is this more apparent than in the religion to which Einstein himself nominally belonged: Judaism. The quest for knowledge is the cardinal sin at the very start, and the very heart, of the Judeo-Christian tradition. In Genesis, the Bible's first book, humankind's original sin is nothing other than *curiosity*. God forbids Adam and Eve to eat from the Tree of Knowledge on practically page one, but the infamous serpent assures Eve that "the day you eat of it your eyes will be opened."[61] The fruit "desirable for gaining wisdom" proves irresistible, and after eating from the Tree of Knowledge Adam and Eve are endowed with a startling new self-awareness.[62] But God isn't pleased with their new knowledge. "Behold," He says to the angels, "the man has become like one of Us, to know good and evil," and He worries that next we'll eat from the Tree of Life and "live forever" like the angels.[63] Blessed with both immortality and self-awareness, we'd be practically gods ourselves. In order to avoid this outcome, human beings are banished forever from the Garden of Eden—and from God's good graces.

It's hard to exaggerate just how damaging and demeaning this myth really is. You'd think that a benevolent God would applaud our curiosity about creation and our interest in enhancing our awareness of ourselves. As Einstein's hero Galileo had declared, "I do not feel obliged to believe that the same God who has endowed us with sense, reason and intellect has intended us to forgo their use."[64] And as Einstein himself once said, "The world consists of real objects, and there are consistent laws underlying them. If we want to honor God, then let us use our reason and intellect to grasp these laws."[65] But for the biblical God, asking questions and heightening awareness weren't just discouraged; they were *crimes*. So it's not surprising that St. Augustine called curiosity "a disease."[66] Nor is it any wonder that Tertullian, the founder of Christian theology, could say in all seriousness that "after Christ Jesus, we have no need of curiosity; after the Gospel, no need of inquiry."[67]

By no means is this condemnation of curiosity limited to Judaism and Christianity. An analogous attitude pervades almost all the world's major faiths. Even the Buddha—often portrayed in the West as a paragon of rationality and open inquiry—discouraged what he considered idle curiosity.[68] The Buddha's teaching was primarily pragmatic, aimed at alleviating the suffering inherent to earthly existence, so he disapproved of most metaphysical claptrap. In one famous parable, a monk threatens to renounce Buddhism if he doesn't get some basic answers to the big questions about the nature of the cosmos and the fate of the soul. The Buddha compares him to a man struck with a poisoned arrow who insists on knowing who shot the arrow, what their name is, and various other irrelevant details, rather than just ending his suffering by having a doctor remove the arrow.

The parable of the poisoned arrow clearly comes from a place of compassion: the Buddha presents himself as the physician who will relieve our suffering, even if he won't answer our deepest questions. But the moral of the story—acquiring knowledge is nothing next to attaining nirvana—is by no means unobjectionable. And although the Buddha's message is a lot less pernicious than the Bible's (in Buddhism, curiosity is merely pointless, rather than punishable), any doctrine that declines to even *try* to answer the deepest questions could never satisfy truly curious spirits.

Einstein preferred the poisoned arrow. Since he saw "the pursuit of truth" as "sacred," pain and discomfort were small prices to pay along the way.[69] "To make a goal of comfort or happiness has never appealed to me," he wrote in his personal manifesto. "A system of ethics built on this basis would be sufficient only for a herd of cattle."[70] But all told, the old religions wanted exactly that: well-behaved believers with subservient souls. Einstein would have none of it. "I believe in the versatility of the human mind," he once said, "and stand for its free development."[71] When a colleague complained to him that "the problem of youth all over the world" was that they had "lost faith," Einstein retorted: "Why shouldn't they doubt? I doubted everything they told me. I wanted to find out for myself."[72]

Since his aim was to foster free, awakened human beings—rather than to fashion yet another rigid religious orthodoxy—Einstein considered curiosity to be both an inalienable right and an immense responsibility. "I believe that the dignity of man depends not on his membership in a church," he said, "but on his scrutinizing mind, his confidence in his intellect, his figuring things out for himself."[73] For the cosmic religion, curiosity was neither idle (as in Buddhism) nor immoral (as in most monotheistic traditions)—it was an admirable, intrinsic impulse toward understanding.

Part of what made curiosity so wonderful was that it *worked*. "The main source of all technological achievements," Einstein argued, "is the divine curiosity and playful drive of the tinkering and thoughtful researcher."[74] The passionate pursuit of questions often revealed real answers—and the resulting knowledge had raised humankind to the highest rank. But for Einstein, answers and achievements were not the main aim; they were an ancillary benefit. "It is not the *result* of scientific research that ennobles humans and enriches their nature," Einstein mused, "but the *struggle to understand* while performing creative and open-minded intellectual work."[75] Whereas received religions always tried to provide reassuring answers, for Einstein it was all about asking revealing questions. "My scientific work," Einstein insisted, "is motivated by an irresistible longing to understand the secrets of nature and by no other feelings."[76] And since deep questioning was practically a sacred duty, neither the direction nor the destination really mattered. "Don't think about why you question, simply don't stop questioning," Einstein once advised a young admirer. "Curiosity is its own reason.... Try to comprehend a little more each day. Have holy curiosity."[77]

Fortunately for him (and for the rest of humankind), Einstein was born into a secular family that fostered his precocious curiosity. At home, he was free from religious dogma. But Germany's militaristic school system stepped in to substitute for the synagogue. "I felt that my thirst for knowledge was being strangled by my teachers," Einstein recalled of his schoolboy days.[78] Einstein saw education, like science, as essentially a spiritual pursuit. He thought schools should be filling young minds with wonder rather than facts—"not knowledge per se," as he put it, "but a *longing* for knowledge and understanding."[79] To him, the real aim of education was to help young minds embark on the age-old quest to play a conscious part in the colorful pageant of Infinity. "The highest goal for any educator is to open the minds of his students," Einstein said. "We are so small, and the universe is so immense. To make a student conscious that he uses only about ten percent of his mental potential, that he is a part of this immensity and can comprehend it if he wants to; this is the duty of any teacher who's worth his salt."[80] Of course, educational systems tended to fail miserably in this task. "It is in fact nothing short of a miracle," Einstein thought, "that modern methods of instruction have not yet entirely strangled the holy curiosity of inquiry; for this delicate little plant, aside from stimulation, stands mainly in need of freedom; without this it goes to wrack and ruin without fail."[81]

Carefully cultivated, however, curious little plants could become towering trees—beings who shaded, sheltered, and served the human community. Still, it would be "a very grave mistake," Einstein thought, "to think that the joy of seeing and searching can be promoted by means of coercion and a sense of duty."[82] Creative individuals—those filled with wonder for the world's grandeur and an insatiable curiosity about its complex inner workings—could flourish only when free from compulsion. For students, this meant freedom from rote memorization and endless examinations;[83] for the citizen, freedom from state propaganda;[84] for the scientist, freedom of inquiry;[85] and for the spiritual seeker, freedom from ossified ideologies that suffocated the spirit. As Einstein saw it, "valuable achievement can sprout from human society only when it is sufficiently loosened to make possible the free development of an individual's abilities."[86]

He considered such "creative and impressionable individuals" to be what was "truly valuable in our bustle of life."[87] These curious minds had created all that was "noble and sublime" in art and science, ethics and religion.[88] And over the centuries, the deep feelings of artists, the high philosophy of mystics, and the sophisticated theories of the scientifically inclined had all shed their own sort of light on our mysterious world. Yet wonder alone had never been enough to grasp the hidden order behind all the apparent chaos of our complex cosmos. Awe awaited some organizing principle; curiosity required some constraints. There had to be a method to the marveling.

Thousands of years ago, some strange intellectuals in southern Italy thought they'd figured out what the mind had been missing. We remember them as the Pythagoreans, and these mystical proto-mathematicians thought they'd found the mental medium that would finally allow us to hear the world's hidden harmony. Over the ensuing centuries, the Pythagoreans died off and their schools were destroyed. But their sacred tradition of searching for a science of the sublime survived. Many great minds achieved much with mathematics, but no one, not even Isaac Newton, had shown the world the method's true majesty. For more than two millennia mathematics would lie waiting, like the sword in the stone, for someone worthy to wield it. And in 1905, Albert Einstein would show the world just what one wonder-struck, curious mind was really capable of.

4

THE HIDDEN HARMONY

I hold it true that pure thought can grasp reality, as the
ancients dreamed.

—Albert Einstein in a lecture at Oxford, June 1933

THE CONCEPT OF RELATIVITY CAME SUDDENLY TO EINSTEIN ONE morning, when he tells us that "a storm broke loose in my mind."[1] From that moment on, he spent every spare second putting his ideas to paper. And although his rational mind was working overtime to elaborate on the initial inspiration, Einstein insisted that in the beginning "ideas come from God." He meant this metaphorically, of course.[2] But as one biographer has pointed out, the end result, his paper on the special theory of relativity, "was strangely free of footnotes or references, as if the inspiration had indeed come, if not from God, from some otherworldly source."[3]

And special relativity was only the climax of this frenzied period of almost ferocious creativity. The same year, Einstein established the existence of atoms, revolutionized our understanding of light, and demonstrated the equivalence of matter and energy.[4] All this he achieved with little more than his own mind—without doing a single experiment, without an academic position, and while working full-time at the patent office in Bern, Switzerland.[5] What astonishing arrogance, what wild ambition, what crazy confidence could have convinced a twenty-six-year-old patent clerk with a diploma in physics under one arm and a baby in the other, pipe in mouth and pen in hand, that he could unravel the enigmas of eternity in his spare time? And even more mind-boggling than the attempt itself is the result: how could he have been *right*?

The story of Einstein's seminal discoveries has been told many times. Every imaginable influence has been meticulously traced, and it's widely agreed that Hume, Mach, Maxwell, and Lorentz were the keys. No doubt they provided the immediate inspiration, and to be sure, these thinkers furnished the intellectual tools needed for the task. But intellectual influences alone aren't enough to explain the spirit that animated the search. Einstein was adamant that his science had always been "guided not by the pressure from behind of experimental facts, but by the attraction in front from mathematical simplicity."[6] He was certain that a grand synthesis of great beauty lay shimmering just beyond the horizon of our current understanding.

He was hardly the first person to be seduced by such a vision. All the intellectual giants upon whose shoulders Einstein stood had strained to see the same thing. And they all shared his faith that the simplicity at the center of all things could someday be seen, because all of them were inheritors of the same sacred tradition handed down for centuries. Most people have never heard of the mysterious man who stands at the start of this tradition, apart from his famous theorem. But his insight that numbers were the right way to reveal reality played a key role in initiating Einstein's quest to understand our cosmos. And with all the magnificent progress Einstein made through the power of pure thought, he himself would prove to be the pinnacle of the Pythagorean tradition. But who was the strange sage who seems to have sowed the seeds of the Scientific Revolution so very long ago?

Knowledge of Pythagoras is patchy, but the renown in which he was held in the ancient world is unmistakable.[7] The historian Herodotus ranked him "one of the greatest Greek teachers."[8] The philosopher Empedocles said he was "a man who had won the utmost wealth of wisdom."[9] And some simply called him "the divine Pythagoras."[10]

He was born sometime around 570 B.C.E. on the island of Samos, off the coast of what's now Turkey.[11] At the time, this coastal region, called Ionia, was filled with Greek colonies and well on its way to becoming the first serious center of Western philosophy. An immense trade network flourished in the area, and Samos became a center for ships sailing all over the Mediterranean. Archaeologists have in fact found no other Greek site so abundant in foreign materials, and this widespread trade brought not just foreign goods, but foreign ways of thinking.[12] With Babylon, Egypt, and Persia all in the immediate neighborhood, Ionia was immersed in what one scholar called "the full current of Oriental thought."[13] The mingling of so many different ideas led to a new skepticism about received religion and new notions of the origin and nature of the world.[14]

Into this mind-expanding milieu, Pythagoras was born. Accounts of his life are semi-legendary and were written centuries after the fact, so they need to be consulted with caution. But biographers agree that Pythagoras had far too refined a mind to remain confined to the tiny island of Samos.[15] Tradition has it that he left home at eighteen and set out on a self-guided pilgrimage of the ancient world, studying with a plethora of priests, philosophers, and scientists around the Mediterranean.[16]

First he ingratiated himself with the greatest Greek philosophers of the age in the nearby city of Miletus, on the Ionian coast. Here he's said to have studied under the philosopher Thales (the same one who was once so lost in thought he fell down a well).[17] After Pythagoras had learned all he could in Miletus, he traveled through the Near East, then spent some twenty years in Egypt. Along the way he absorbed all the eclectic wisdom that would come to define his school: mathematics and geometry, vegetarianism and reincarnation, meditation and memory techniques.

Already in ancient times, the philosopher Heraclitus was heaping scorn on him for all his allegedly superficial studies.[18] But this scorn only shows that everyone, even his enemies, acknowledged that Pythagoras was a polymath.

After decades studying abroad, Pythagoras returned home to found his first philosophical school on Samos. But with the rise of the tyrant Polycrates, Pythagoras figured he'd be better off somewhere a little less hostile to freethinkers, and he set sail for the still-tolerant shores of southern Italy. Known as Magna Graecia (Greater Greece), these close-knit but fiercely competitive colonies were almost as much of a cultural crossroads as the Ionian coast.[19] Pythagoras picked the city of Croton for his new community, and it proved to be a propitious choice. The Crotonians welcomed this wise, warmhearted man with open arms, and he soon attracted followers in the thousands. His students became well known for simple living, shared possessions, and a passionate quest for both scientific and spiritual truth. The central axiom of their school was simple: our cosmos was held together by a hidden harmony, and mathematics was the best means of understanding it.[20]

Pythagoras is remembered and respected for many reasons. He was the first to call himself a *philosopher* (literally, a "lover of wisdom"). He's also said to have coined the word *cosmos*, which in ancient times implied a beautiful and orderly arrangement of the world.[21] But like the persistent myth that Einstein was a naïve bumbler, "the usual stereotype" of the Pythagoreans, according to one scholar, is that they were "impractical dreamers, their minds fogged and obsessed with number mysticism, who had no clear idea of the value of empirical research because all that interested them was discovering metaphysical principles."[22] And while they certainly seem to have embraced some odd ideas, there was much more to them than just a reverence for numbers or a revulsion for beans.

Ancient authors tell us that Pythagoras was the first to say "that the earth is spherical," and also the first to realize that the morning star and evening star were one and the same: the planet Venus.[23] By around 400 B.C.E., the Pythagorean Philolaus was championing a heliocentric model of the solar system and suggesting that the moon and stars were other worlds that might be inhabited.[24]

The human body interested the Pythagoreans just as much as the heavens. Pythagoras himself is said to have realized that "reason and mind are located in the brain"—which makes sense considering that the finest and most famous medical school in the ancient world was located in Croton, the same city where he'd founded his philosophical community.[25] Here, almost a century before Hippocrates, diseases of the human body were studied scientifically by famous doctors like Democedes and Alcmaeon (both Pythagoreans).[26] Alcmaeon even traced the optic nerve from the eye back into the brain and realized that here was where sensations were combined to create perception, thought, and memory.[27]

And as if being at the forefront of mathematics, astronomy, and medicine weren't enough, classicist Peter Kingsley contends that "the one person who during this period was associated most closely with applied mathematics and, in particular, with breakthroughs in the field of mechanics, was none other than the famous Pythagorean from Tarentum: Archytas."[28]

Yet for all their prowess in the practical realm, the Pythagoreans were undeniably a religious group. Sophisticated science was only one part of the Pythagorean life; the other was passionate spiritual practice. Pythagoras himself underwent intensive spiritual training with several different schools. The first tale we're told is of a lengthy spiritual retreat on Mount Carmel (in modern-day Israel), "where, in the temple on the peak, Pythagoras for the most part had dwelt in solitude." Carmel was already known then as "more sacred than other mountains," and it had a long history as a holy place for the Jewish people.[29]

Whatever Pythagoras practiced there, it had a profound effect. By the time he hiked down to the coast to catch a ship headed for Egypt, he was already an ardent spiritual practitioner "whose body had become emaciated through the severity of so long a fast." The pirates who picked him up had planned to sell him into slavery, but we're told that "during the whole trip, [Pythagoras] sat silent where he would be least likely to inconvenience them." And he wasn't just sitting quietly in a corner. In one of the most remarkable (yet least discussed) passages in all of ancient literature, Pythagoras is presented practicing what is unmistakably seated meditation. "For two nights and three days," says the ancient author Iamblichus, "Pythagoras had remained in the same unmoved position, without food, drink, or sleep, except that, unnoticed by the sailors, he might have dozed while sitting upright."[30] Stunned by his strange spiritual practices, the sailors felt that "a divine guardian spirit had crossed with them" over the sea, and they deposited Pythagoras safely on the Egyptian coast.[31]

After arriving in the land of the pharaohs, Pythagoras "visited all of the Egyptian priests, acquiring all the wisdom each possessed. He thus passed twenty-two years in the sanctuaries of temples," Iamblichus says, "being initiated in no casual or superficial manner in all the mysteries of the gods."[32] When he returned home and founded his first school on Samos, "outside of the city he fashioned a cave, adapted to the practices of his philosophy, in which he spent the greater part of day and night, ever busied with scientific research, and meditating."[33] On the island of Crete, he's said to have spent twenty-seven days underground engaged in religious practices in a sacred cave famous throughout the Greek world.[34] And after he founded his second school, in Croton, he again built an underground chamber for the purpose of engaging in spiritual practices.[35]

We don't need to give every one of these stories credence to see that intensive spiritual practice played a pivotal role in Pythagoras's life. We don't know exactly what he was practicing in all these cave retreats and temple sanctuaries, but many scholars over the centuries have suspected the influence of the East. The ancient author Apuleius alleged that Pythagoras traveled as far as India and was instructed in person by Eastern sages.[36] Schopenhauer surmised an indirect Indian influence via Egypt.[37] And the modern scholar Thomas McEvilley has suggested that Jainism—which emphasizes ascetic spiritual practice and places the highest value on vegetarianism and nonviolence—might have had an indirect influence on the Pythagoreans.[38] Finally, there's a strange tale about Pythagoras being visited by a spiritual emissary named Abaris, from an unspecified but faraway land. The eminent classicist Peter Kingsley is convinced that Abaris was a Mongol shaman, and as crazy as it sounds, he makes a compelling case.[39]

But in the end, it doesn't really matter whether or not Pythagoras was influenced by the East. Much more important than who taught Pythagoras is what Pythagoras taught. And while the specific spiritual practices of Pythagoras himself, and their origin, will most likely remain a mystery, we do know a decent amount about what he taught his disciples. His students were immersed in what can best be described as a comprehensive scientific-spiritual educational program, a rigorous discipline designed to lead to the harmonious development of the whole human being.

First off, they were expected to spend *five years* in total silence before they were allowed to absorb the deeper teachings.[40] When they were finally allowed to open their mouths again, Pythagoras advised them "above all things to speak the truth, for this alone deifies men."[41] Disciples were admonished to "devote themselves diligently to learning," and "they took it for granted that studies and disciplines implied labor, and that they must expect severe tests of different kinds."[42] Emotional equanimity was to be maintained at all times, in the face of both good and bad fortune.[43]

The Pythagoreans also "made it a point to exercise their memories systematically, considering that the ability of remembering was most important for experience, science, and wisdom."[44] They even turned the art of memory into a kind of ethico-religious practice. Pythagoras said that "special regard should be given to two times of the day: the one when we go to sleep, and the other when we awake. At each of these we should consider our past actions, and those that are to come."[45] For his students, this "served the double purpose of strengthening the memory and considering their conduct."[46] Reincarnation (*metempsychosis* to the Greeks) was also a central feature of the teaching, but this was no idle belief.[47] It implied the wider insight that "all animated beings are kin...and should be considered as belonging to one great family."[48] Pythagoreans were therefore "ordered to abstain from all animal food" and enjoined to practice a "universal amity of all towards all."[49]

And while "the object of these disciplines had been to turn out good and honest men," Pythagoras also thought that "the purified mind should be applied to the discovery of beneficial things."[50] For the disciples of Pythagoras, development of character and dedication to inquiry were intertwined, even inseparable. And it's not hard to see how the Pythagoreans' spirituality could have complemented and encouraged their science. It's obvious that a love of learning, an insistence on speaking the truth, and a no-nonsense assessment of one's faults and virtues might foster mental qualities useful for a proto-scientist. It's also easy to see how a reverence for life could encourage research in medicine, or how a worship of mathematics and geometry might lead to breakthroughs in mechanics and engineering. The rigorous Pythagorean training regimen was not simply a matter of masochism, then, or religious fanaticism. As one expert on the Pythagoreans has explained this program, "Not only did they study *harmonia* as a universal principle, seeing it reflected on all levels of the beautiful cosmos, they incorporated the principle into the fabric of their daily lives as well."[51] Pythagoras was trying to prepare his pupils to perceive the deepest structures of the cosmos—and trying to teach them to put the principles of the macrocosm into practice at the microcosmic level of individual existence.

Not long after his numinous experience with the compass needle, Einstein encountered the profound school of thought founded by Pythagoras when his uncle Jakob introduced him to the famous Pythagorean theorem.[52] Einstein became obsessed, and "after much effort," he later recalled, "I succeeded in 'proving' this theorem."[53] Completing the proof was a revelation. "I was thrilled to see that it was possible to find out truth by reasoning alone, without the help of any outside experience," Einstein remembered. "I became more and more convinced that nature could be understood as a relatively simple mathematical structure."[54]

Einstein was beginning to embrace what could be called the fundamental Pythagorean faith: we inhabit an orderly cosmos comprehensible with mathematics. This notion soon suffused both his spirituality and his science. He came to believe that "a conviction akin to a religious feeling, of the rationality or intelligibility of the world, lies behind all scientific work of a higher order."[55] And on the other hand, he felt that "whenever this feeling is absent, science degenerates into uninspired empiricism"—a lifeless accumulation of facts that could never result in real understanding.[56] Wonder might have been the emotional wellspring of Einstein's spirituality, but the Pythagorean faith was its firm intellectual foundation. And even though he was forthright about who fathered his belief that "true scientific thought is not possible without faith in the inner harmony of our universe," no one has given proper pride of place to Pythagoras as the principal source of the only axiom Einstein ever accepted without evidence.[57]

This oversight isn't surprising; the influence of the Pythagoreans in general is grossly underrated. The concept of a harmonious underlying order is, of course, found in many cultures. From the cycling of the seasons to the phases of the moon, it's impossible not to recognize regularity in the world around us. But the idea that this orderly harmony is mathematical in nature is usually considered Pythagorean.[58] And although it's often forgotten, the fundamentals pinpointed by the Pythagoreans still form the foundation of our worldview today. In her masterly history of how this seemingly simple notion has shaped Western science for more than two millennia, historian Kitty Ferguson highlights how the Pythagoreans' "confidence that truth is accessible by way of numbers" and their "assumption that there is unity to the universe" became "the pillars undergirding science."[59] Especially in Einstein's field, physics. "It is the modern-day physicists who have come most close to approximating Pythagorean conceptions," notes David Fideler, another expert on Pythagoras. "The deeper they push into matter the more it looks like the cosmos of the Pythagoreans. Each atom is a Pythagorean universe, the sight of eternity in a grain of sand."[60]

The founders of quantum physics were well aware that they were carrying on this ancient tradition. Erwin Schrödinger thought that Pythagoras "must have been one of the most remarkable persons of antiquity," and called the Pythagorean faith in mathematics "a sweeping generalization of truly imposing boldness and grandeur."[61] Werner Heisenberg, also well acquainted with the ancient Greeks, agreed that "the Pythagoreans seem to have been the first to realize the creative force inherent in mathematical formulations."[62] And Wolfgang Pauli noted the outsized influence of Pythagoras on all subsequent science and philosophy.[63] Even Einstein's favorite atheist, Bertrand Russell, agreed. "Mathematics, in the sense of demonstrative deductive argument, begins with him," he wrote.[64] Although familiar with thousands of thinkers, Russell ranked Pythagoras above them all: "I do not know of any other man who has been as influential as he was in the sphere of thought."[65]

His influence isn't hard to trace. Copernicus (1473–1543) quoted ancient Pythagorean authorities in support of his heliocentric system,[66] and according to one historian he "fancied himself the restorer of ancient wisdom [and] insisted that his system was no rash innovation, but a revival of...the lost doctrine of Pythagoras."[67] Tycho Brahe (1546–1601) had his palace/observatory, Uraniborg, built according to perfect Pythagorean proportions, and within its walls he collected the careful astronomical observations that later permitted Kepler to discover the laws of planetary motion.[68] Galileo (1564–1642) acknowledged that he was advocating the "Pythagorean opinion that the earth moves" on the very first page of his famous *Dialogue* (the book that would soon be banned and get him branded a heretic).[69] And no less a mind than Isaac Newton (1643–1727) was convinced that his theory of universal gravitation exemplified the Pythagorean ideals of simplicity and unity.[70]

Although Einstein had sensed the "beginnings" of his third phase in Buddhism and some of the biblical prophets, he saw the full flowering of his cosmic religious feeling among the founders of modern physics.[71] He considered people like Kepler and Newton exquisite exemplars of his cosmic religion: pioneering scientific minds with soaring spirits to match.[72] "What a deep conviction of the rationality of the universe and what a yearning to understand!" Einstein exclaimed.[73] Kepler in particular was Einstein's personal idol. "There we meet a finely sensitive person," he said, "passionately dedicated to the search for a deeper insight into the essence of natural events, who, despite internal and external difficulties, reached his loftily placed goal."[74] What impressed Einstein wasn't just Kepler's triumphant success, but his tenacity and stamina. "Neither by poverty, nor by incomprehension of the contemporaries who ruled over his life and work did he allow himself to be crippled or discouraged," Einstein enthused.[75]

Still, even Kepler wasn't quite there. Although he was well aware of his intellectual ancestry (he called Pythagoras the "grandfather" of modern cosmology), for him there could be no family feud between scientific grandfather and heavenly Father, no contradiction between his Pythagorean and Christian faiths.[76] Kepler figured that the two actually complemented each other: a benevolent God would naturally create an orderly cosmos and grant human beings the intelligence necessary to grasp His creation. According to Ferguson, "This was the conviction that set fire to [Kepler's] spiritual and scientific imagination, and that flame would last him a lifetime."[77]

The same flame illuminated Einstein's mind, but there was a fundamental difference. Kepler kept his Christianity, although he admitted that "God also wants to be known through the Book of Nature."[78] Einstein could never accept such a compromise. He dismissed religious writings altogether as a source of truth and argued that God could be known *only* through the book of Nature (not *also*, as Kepler contended).[79] "The scientists of tomorrow," Einstein said, "with ideas as infinite as the universe, will no longer read fairytales about creation, but only the one authentic book written by creation itself. They will discover through mathematical computation laws governing the universe."[80] It would be hard to pen a more pithy epitome of the Pythagorean faith.

Once we acknowledge that all the major physicists of the past five hundred years have been enchanted with the Pythagoreans, it becomes easy to understand how Einstein could have had such a strong conviction that our cosmos was comprehensible even though he didn't believe in a personal God. And yet the pious myth persists that science is somehow the offspring of monotheism—that the Judeo-Christian tradition was what gave history's greatest scientists confidence in the rational order of reality. This outrageous presumption is known as the dependency thesis (as in, science *depends* on Christianity). The central claim is that "without Christianity, modern science could not exist."[81]

Far from a fringe theory pushed only by biased theologians, the thesis has been promoted by many intellectual heavyweights. Philosopher Alfred North Whitehead thought that "faith in the possibility of science" was an "unconscious derivative of medieval Christian theology."[82] Historian Joseph Needham argued that only the belief in a personal God could have given us the notion of an orderly cosmos for science to study.[83] And Harvard biologist E. O. Wilson "accepts it as a given."[84]

But even in principle, the conceit that Christianity encourages rational thinking, much less rigorous science, is ridiculous. A personal God, practically by definition, is one who *suspends* physical law, rather than sustains it—and from Moses parting the Red Sea to Jesus turning water to wine, the Bible boasts a multitude of miracles.[85] God is always ready to violate Nature's laws to reward (or, more often, punish) His creatures.[86] And even if miracles could somehow be squared with science—as claimed by famous figures like Francis Collins, who led the Human Genome Project—on balance we'd all be better off without them.[87] The occasional resurrection really can't outweigh all the locusts and leprosy, the plagues and demon-donkeys, the people getting turned into pillars of salt or torn to pieces by God-sent bears (yes, that).[88] To be fair, the God of the New Testament cures a lot more leprosy than He causes, but that's hardly evidence that Christianity inspires rational thinking.

And if the dependency thesis is untenable in principle, it's downright absurd in practice. Never mind the fact that mathematics wasn't born in the Western world, but was borrowed from Mesopotamia, Egypt, and India.[89] Or that many seminal innovations derived from our supposedly Christian science were first made in China (the compass, gunpowder, paper, and the printing press are the classic examples). Even if we focus only on the Western world, it's a curious coincidence that science was in full bloom for centuries before Christ came along, but then languished for more than a millennium after he arrived.[90]

Remember the Dark Ages? Historians are right to insist that *some* progress was certainly made during this time, and that the poor people who populated this benighted era weren't quite as obtuse as we tend to think. But even granting that "the 'Dark Ages' weren't as dark as we thought," that doesn't exactly make them dazzling.[91] There's no escaping some simple facts: the influential discoveries and incredible inventions in the centuries before the birth of Christ could fill a whole book—and they have.[92] But by the time we reach the depths of the Dark Ages, historians are hailing the horseshoe as a major technological achievement and giving us the Gothic cathedral as an example of breathtaking innovation.[93]

Gothic cathedrals are great, of course, and the horseshoe was indeed an important invention. But for all the importance of beautiful buildings and practical technological improvements, neither art nor technology is *science*. Neither depends on—or gave rise to, or encouraged—a systematic and critical approach to refining and unifying our knowledge of the world. And nor did Christianity. Call it a coincidence, call it a misleading myth, call it the Middle Ages—no matter how you finesse it, the zenith of Christianity was the nadir of intellectual innovation in the Western world.[94] And yet, after being steeped for more than two thousand years in the Judeo-Christian tradition, we can hardly even imagine any other source for our science.

But of course, an equally important source of Western civilization was ancient Greece, an inheritance best exemplified by the philosophy of Plato (c. 428–348 B.C.E.). Plato was prolific: his complete works are comparable in length to the Bible, and he remains rightly revered as one of the greatest and most original of Western thinkers.[95] Alfred North Whitehead famously said that "the European philosophical tradition" was only "a series of footnotes to Plato."[96] But it's no insult to Plato's perspicacity to point out that he too was the inheritor of an ancient tradition. And if we dig down even a little below the surface, the Pythagorean roots of many of his ideas are soon revealed.

No less an authority than Aristotle—Plato's star pupil, who spent twenty years studying at the Platonic Academy—wrote that "the philosophy of Plato . . . in most respects followed these thinkers."[97] Specifically, he pointed out that Plato "agreed with the Pythagoreans in saying that the One is substance and not a predicate of something else; and in saying that the numbers are the causes of the substance of other things."[98] In other words, Plato accepted the two key teachings embraced by Pythagoras a little before him and Albert Einstein long after: that all is ultimately One, and that mathematics is both an expression of and a means of access to the eternal order underlying all things.

Aristotle was by no means the only one who believed his mentor was both an inheritor and an embodiment of Pythagorean wisdom. This perspective on Plato endured all through antiquity. Some seven hundred years after Aristotle, the Neoplatonist philosopher Proclus was still arguing that Plato had imbibed "an all-perfect science of the divinities from the Pythagoric and Orphic writings."[99]

This is hardly just a hypothesis. It's well known that after the execution of Socrates, Plato left Athens to travel to Magna Graecia to study with the Pythagoreans. In the city of Tarentum, on the heel of Italy's boot, he entered into a sacred ritual relationship as the honored guest and student of the Pythagorean patriarch Archytas.[100] Archytas of Tarentum (c. 435–360 B.C.E.) was famous throughout the Mediterranean as a master mathematician and mechanical engineer, as well as the military leader of one of the most powerful city-states in the ancient Greek world. And Archytas is said to have studied with Philolaus (c. 470–385 B.C.E.), the most important Pythagorean in the generation just after Pythagoras.[101] So it seems that Plato didn't just read some Pythagorean treatises or talk to a few Pythagoreans who happened to be visiting Athens. He went out of his way to go and study, in person, with the most famous Pythagorean of the day—a man who could claim a direct line of transmission of esoteric oral teachings going all the way back to Pythagoras himself.[102]

Although it's frequently forgotten (or deliberately ignored) today, Plato often alluded to his enormous debt to the Pythagoreans. There are many dialogues in which Plato takes no credit for key ideas he puts into Socrates's mouth, instead admitting that they originate in oral traditions transmitted to him by other teachers. It's clear from various contextual clues that the wise mentors Plato so often mentions must be the Pythagoreans he studied with in southern Italy.[103] And even after returning to

Athens, Plato was still in direct contact with Pythagoreans on the other side of the Ionian Sea. We even have a letter he wrote in which he thanks Archytas for sending him more Pythagorean treatises to study.[104]

And then there are the myths. Plato is remembered mainly for enshrining the idea that philosophy is a dialectical process in which clear logic, sharp definitions, and eloquent argument are employed to approach ever closer to what's eternally true. This model of philosophy as an eminently rational enterprise is essentially how modern philosophy still sees itself today. But many of Plato's most famous dialogues contain—or even conclude with—mind-bending myths about the creation of the cosmos, the immortality of the individual soul, and the flight of the philosopher to celestial realms upon spiritual wings.[105] These myths were obviously chosen with care as essential adjuncts to the rational discussions that make up the bulk of the dialogues: they complement the intellectual content by offering evocative imagery and an emotional impact that clever arguments alone could never convey. Through some deft detective work, classical scholars have been able to show beyond doubt that many of these myths come from Sicily and southern Italy—the strongholds of Pythagoreanism in Plato's time, and the very places Plato himself traveled to repeatedly.[106]

Nowhere is the influence of the Pythagorean tradition more transparent than in the *Timaeus*. The *Timaeus* is one of Plato's last and greatest works, a majestic monument to the mature thought of his later years.[107] The eponymous hero of the dialogue, Timaeus, is described as a kind of philosopher-king: both the foremost thinker and the political leader of the city of Locri in southern Italy. Ever since ancient times it's been suspected that Plato modeled this fictional character on a real Pythagorean, probably either Philolaus or his own teacher, Archytas.[108]

The teachings of the *Timaeus* seem to confirm this suspicion. The core of the dialogue is a creation myth in which Timaeus tells us that our cosmos is "a work of craft, modeled after that which is changeless."[109] According to Timaeus, a "supremely good" geometer-god he calls the *demiourgos* "constructed the visible and tangible universe" and "brought it from a state of disorder to one of order," in the process "making it a symphony of proportion."[110] He goes on to offer an elaborate explanation of how our universe is actually an agglomeration of innumerable tiny triangles and tetrahedrons.[111] When a prodigious number of these "unalloyed primary bodies" are "clustered together" and made "commensurable and proportionate" by the divine intelligence of the *demiourgos*, they give rise to our unified, harmonious cosmos.[112]

He applies the same logic to human beings. Timaeus saw the human soul as an amalgam of different elements integrated into a harmonious whole by perfect laws of proportion.[113] In this way, he considered the "combination of soul and body which we call the living thing" as an image of the Eternal, an "imitation of the structure of the universe."[114] And for Timaeus, this "affinity to the divine" not only explained our origins; it also indicated the way to enlightenment.[115] "By coming to learn the harmonies and revolutions of the universe," he claimed, we could bring ourselves back "into conformity" with the cosmic order. "And when this conformity is complete," he believed, "we shall have achieved our goal: that most excellent life offered to humankind."[116]

Here in the *Timaeus*, some 350 years before the birth of Christ, Plato had already spelled out the basic Pythagorean belief that our entire cosmos, and even our own souls, had been constructed according to rational rules and perfect proportions by a kind of divine intelligence. And he'd also articulated the equally important idea of math as a mystical path that could lead us to direct contact with the divine reality.

And yet the *Timaeus* provides only one specific example of a much more general phenomenon: Pythagorean ideas can be found everywhere throughout the work of the most well-known and well-read philosopher in the entire Western canon. It would be an exaggeration to simply call Plato a Pythagorean, but it's probably fair to say that the influence of the Pythagoreans was second in importance only to that of Socrates. In the words of one classicist, "Plato himself is not so much a starting-point as an isthmus between two continuous bodies of tradition"—a crucial link in the chain connecting Presocratic thinkers like Pythagoras to the subsequent intellectual tradition that ultimately evolved into modern philosophy, mathematics, and physical science.[117]

The defenders of the dependency thesis seem determined to ignore our enormous debt to the Pythagoreans. But even Einstein—no expert on intellectual history—recognized that modern science was a victory over Judeo-Christian culture rather than a result of it. In his hero Galileo, he saw someone who had "the passionate will, the intelligence, and the courage to stand up as the representative of rational thinking against the host of those who, relying on the ignorance of the people and the indolence of teachers in priest's and scholar's garb, maintain...the rigid and authoritarian tradition of the Dark Ages." Galileo's great contribution had been "to overcome the anthropocentric and mythical thinking of his contemporaries and to lead them back to an objective and causal attitude toward the cosmos, an attitude which had become lost to humanity with the decline of Greek culture."[118]

For Einstein, it was "the spirit of ancient Greece" that "challenges the individual to think, observe and create," and he applauded "the great materialists of ancient Greek civilization" for asserting that "all material phenomena" were "controlled by rigid laws."[119] So in Einstein's eyes, it was only right that "we reverence ancient Greece as the cradle of western science."[120] The sages of the Scientific Revolution had ended what Einstein called "the long hibernation of Occidental thought" by reclaiming this ancient attitude of inquisitiveness about the awesome order evident everywhere in Nature.[121]

And once this attitude had been regained, Einstein was adamant that it must be maintained, because "true scientific thought," he felt, "is not possible without faith in the inner harmony of our universe."[122] This wasn't just a broad historical generalization; it was a principle derived from deep personal experience. During both his early years, when he worked in obscurity, and his later years, when he faced failure after failure, this conviction was "a kind of faith that helped me through my whole life."[123] It was an undying guiding light across all "the years of anxious searching in the dark for a truth that one feels but cannot express."[124]

Since Christianity can't take credit for fostering this faith, or all the fabulous things that have followed from it, we should probably look elsewhere—perhaps at the tradition that all the greatest scientists of the past five centuries have always pointed to. Among the Pythagoreans, we'll find an ancient community that maintained a religious reverence for reality while still encouraging free scientific inquiry, a society of seekers who thought that reason and religion supported and supplemented one another. Neither a numerical understanding of Nature nor a numinous feeling of the divine was enough for them. Thousands of years before Einstein conceived of his cosmic religion, the Pythagoreans, too, were seeking a synthesis of science and spirituality.

The Pythagoreans' primary goal was "to achieve a certain divine union," and since they saw number as the secret source of the universe's harmonious structure, they reasoned that math must be the royal road to both rational *and* religious understanding.[125] "The study of mathematics was not a *preparation* for the contemplation of a divine Reality," says classicist Cornelia de Vogel, "*it was the contemplation itself.*"[126] For the Pythagoreans, mathematics was more than just a scientific method—it was a mystical path. This strange synthesis of math and mysticism exemplified "the persistent human longing," as another scholar has called it, "to combine the hypnotic spell of the religious with the certainty of exact knowledge—an ideal which appeals, in ever changing forms, to each successive generation."[127]

Einstein's generation was no exception. His fellow physicists both recognized and respected the Pythagorean search for a synthesis of science and spirituality. Wolfgang Pauli acknowledged that the effort "to effect a synthesis of the basic attitudes of science and of mysticism" in fact "originates with Pythagoras."[128] Werner Heisenberg realized that "here has been established the connection between religion and mathematics which ever since has exerted the strongest influence on human thought."[129] And Bertrand Russell believed that within this school there was "an intimate blending of religion and reasoning, of moral aspiration with logical admiration of what is timeless, which comes from Pythagoras."[130]

Einstein was the very epitome of this ancient effort to amalgamate reason and religion. "Among the profounder sort of scientific minds," he

80

insisted, the "religious feeling takes the form of a rapturous amazement at the harmony of natural law."[131] For Einstein, faith in this harmonious order was an essential factor in advancing our understanding. "Without the belief in the inner harmony of our world," he thought, "there could be no science."[132] And he freely admitted that his own arduous search for a unified theory in physics had been sustained by this Pythagorean faith. "If I hadn't an absolute faith in the harmony of creation," he once said, "I wouldn't have tried for thirty years to express it in a mathematical formula."[133] The essence of this approach was a search for simplicity: the Pythagoreans had spoken of seeing a simple Oneness at the center of existence, and Einstein agreed that "the simpler our picture of the external world, the stronger it reflects in our minds the harmony of the universe."[134]

The fact that Einstein kept using the word *harmony* to express his ideal of reality is revealing. Philologists (scholars who study the history of language) realized long ago that the ancient Greek words for harmony and number, ἁρμονία (*harmonia*) and ἀριθμός (*arithmos*), likely derive from the same root: ἁρω (*harō*). Over time, the words diverged and came to have very different meanings for most.[135] But according to classicist Alister Cameron, among the Pythagoreans these twin concepts continued to be intimately intertwined. "There was no religious school in early Greece vitally concerned with Number except the Pythagorean," he points out. "There was no other philosophical school in whose dogma Numbers and Harmonia played such an important part, and, finally, no other school, religious or philosophical, in which Number *was* Harmonia."[136] These were the quintessential concepts for the sages of southern Italy, "the two most significant words in [the] Pythagorean vocabulary."[137] And for Pythagoras and his followers, this notion that reality was held together by a numerical harmony was not just another interesting idea; it was "an attitude toward life, an all-embracing myth."[138]

Einstein adopted an analogous attitude, and he knew he wasn't alone in his adherence to this peculiar Pythagorean faith. Every scientist, he insisted, "insofar as he considers the viewpoint of logical simplicity as an indispensable and effective tool of his research," was in fact a

"Pythagorean."[139] Yet aside from this one quotation, Einstein hardly ever mentioned Pythagoras or his followers.[140] Could he really have known that he stood at the end of an ancient tradition that sought to unify science and spirituality? Could he have been conscious of his debt to the past—or aware that he himself was the pinnacle of the Pythagorean tradition?

The answer lies in Einstein's letters—and his library. Although he only rarely refers to the followers of Pythagoras, he did revere one Pythagorean in particular. Like Pythagoras, this ancient Greek philosopher was born near the epicenter of the intellectual renaissance sweeping through ancient Ionia.[141] He's said to have spent his whole inheritance on books and his whole youth traveling to study with illustrious teachers.[142] He wrote dozens of books, including the first one ever about Pythagoras.[143] No less than Archimedes credited him with being a pioneer of mathematics, and legend has it that Hippocrates once came across him dissecting animals, searching for the biological basis of mental illness.[144] And yet this pioneering scientist was no stranger to the spiritual. "He used to train himself in a great variety of ways," the ancient biographer Diogenes tells us, "sometimes living in solitude and spending time in tombs."[145] Others said that "he goes off sometimes into the Boundless," letting his spirit soar into strange realms.[146] And like Pythagoras, he too descended down into caves for spiritual retreats.[147] Einstein called him a "religious genius," but his spiritual sojourns are mostly forgotten today.[148] History remembers him instead as one of the founders of modern science, the father of atomic theory.

His name was Democritus. And as it turns out, Einstein's lifelong friend Maurice Solovine translated his teachings from ancient Greek and wrote an entire book about him. The book still resides in Einstein's personal library, and in 1930 he wrote Solovine a letter telling him he'd been "elated" after reading it.[149] He wasn't just flattering an old friend. Einstein very rarely made markings in his books, but Solovine's *Democritus* is graced with many of Einstein's own handwritten highlights. One of the passages that caught Einstein's eye explained how the higher kind of human being should feel at home in the whole universe. "To the wise

170.

Toutes les peines sont plus agréables que le repos quand on atteint le but, ou quand on est sûr de l'atteindre. Mais quand on subit un échec, tout effort est également pénible et fatigant.

171.

Garde-toi, même quand tu es seul, de dire ou de faire ce qui est mal, mais apprends à te sentir plus honteux devant toi-même que devant les autres.

172.

Les lois n'empêcheraient personne de vivre à sa guise, si les hommes ne se maltraitaient pas mutuellement. L'envie est, en effet, la source de la discorde.

173.

Le séjour à l'étranger apprend à se suffire à soi-même. Un morceau de pain d'orge et un lit de paille sont les remèdes les plus doux contre la faim et la fatigue.

174.

La terre tout entière est ouverte à l'homme sage, car la patrie d'une âme élevée, c'est l'Univers.

Pictured here is a page from Einstein's personal copy of the sayings of the ancient Greek philosopher Democritus, as translated into French by his lifelong friend Maurice Solovine. Democritus is remembered primarily as a proto-scientist, the father of the atomic theory of matter. But as a follower of the Pythagorean school, he supplemented his scientific pursuits with spiritual practices. So it's no mystery why Einstein referred to him as a "religious genius." The distinctive diagonal pencil stroke over §174 was Einstein's characteristic way of marking a passage of special interest. The passage reads: "To the wise man every land is open, for the homeland of an elevated soul is the whole cosmos." Photo © the author (2023), used with permission of the Albert Einstein Archives.

man every land is open," Democritus declared, "for the homeland of an elevated soul is the whole cosmos."[150] Einstein was also especially impressed with how "beautifully" Solovine had "handled Democritus' relationship to his predecessors."[151] There can be no doubt that Einstein knew who these predecessors were, because Solovine points out on the third page that Democritus was "a zealous follower of the Pythagoreans."[152]

Although Solovine covered the scientific theories and moral maxims of Democritus in detail, he also described him "frequenting deserted places" and "spending time among tombs" in order to engage in spiritual exercises.[153] So it's no mystery why Einstein saw Democritus as a spiritual (and not just scientific) genius. Solovine even spelled out the fundamental Pythagorean faith that Einstein followed his whole life. "A conception amazing in its audacity, and in the incalculable consequences it had in the history of science and philosophy, is that of the Pythagoreans," Solovine enthused. "They proclaimed that 'numbers are the elements of all things' [and] were the first to recognize the role that mathematics played in the knowledge of things."[154]

So it's plain that Einstein was well aware that the Pythagoreans had played a preeminent role in initiating this intellectual adventure. In fact, he later told an interviewer that there was great "significance" in the "beautifully expressed idea...of Pythagoras, which asserts number to be the nature of all things."[155] Naturally enough, Einstein saw himself as both an inheritor and an embodiment of this ancient tradition. When a young friend once asked him how he wanted to be remembered, Einstein listed a long line of great minds who'd sought to know the divine through mathematics and geometry. "Galileo, Copernicus, Ptolemy, Pythagoras, Newton, Aristotle, and myself," he mused. "That, I guess, is how I would like to go down in history."[156]

In a very real sense, all our science, and all our efforts to merge it with the things of the spirit, is a renewal of a quest begun—and interrupted—more than two millennia ago. The seeds sown by the Pythagoreans had long lain dormant, but nourished in the rich soil of the Renaissance they sprouted again in the fifteenth century. Copernicus, Kepler, Galileo, and

Newton were the first green shoots that sprang forth from this fertile ground, but with Einstein the ancient promise of Pythagoras reached its full flowering. Einstein knew the Pythagoreans were his intellectual and spiritual ancestors, and he also must have been aware that he was one of their most dazzling disciples. Never before—and never since—has a human mind fathomed so much by means of mathematics. Recognizing the key role played by Pythagoreanism in this process not only reveals the real origins of Einstein's rational spirituality, but also points us toward the true guiding spirit of the entire scientific enterprise: to know not the greatness of God, but the grandeur of Nature.

Yet for the Pythagoreans, there was no real difference. In their eyes, mathematical understanding was tantamount to "direct contact with a divine Reality," and Einstein saw science the same way.[157] "There comes a point," he said, "where the mind takes a higher plane of knowledge," and he often experienced a very Pythagorean kind of mystical ecstasy in his scientific quest.[158] "The mechanics of discovery are neither logical nor intellectual," he insisted. "It is a sudden illumination, almost a rapture."[159]

Many scholars have tried to understand how Einstein arrived at his extraordinary insights.[160] And while Einstein was also intrigued by the inner workings of his own mind, he was under no illusion that he fully understood the ontogeny of his ideas. "It is not easy to talk about how I reached the idea of the theory of relativity," he once admitted. "There were so many hidden complexities to motivate my thought."[161] But if anything was clear to him after analyzing his own mental machinery, it was that the mysterious process by which his mind grasped Nature was non-verbal, non-rational, non-logical. "When I examine myself and my methods of thought," he said, "I come close to the conclusion that the gift of imagination has meant more to me than my talent for absorbing absolute knowledge."[162]

Einstein called his imaginary intellectual excursions "thought experiments" (*Gedankenexperimente*), and it was via these imaginal exercises that he arrived at the insights that made him immortal. His best-known vision occurred at just sixteen years old, when he imagined himself traveling through the vacuum of space at the same speed as light itself. He soon forged this thought experiment into a formal theory that revolutionized physics: special relativity.[163] Ventures into the visual imagination continued to play a key role in Einstein's creative process throughout his scientific career, and he always had a high estimate of the value of visual thinking.[164] "Imagination is more important than knowledge," he insisted. "For knowledge is limited, whereas imagination embraces the entire world, stimulating progress, giving birth to evolution."[165]

Einstein's ascent to the heavens in search of a transcendent knowledge of Nature is a striking story, often told, and his imaginary explorations might seem like a radical new way of doing science.[166] But actually Einstein's method was merely a modern variation on an ancient theme, a distant echo of a primordial pattern present everywhere in the world's religions. Every spiritual tradition places immense value on visionary experience as a pathway to knowledge. An imaginal ascent beyond the sky and the stars to a celestial realm where the heart of reality is revealed is in fact the most ancient motif in the history of the human spirit. The oldest written records of religious experience, the hieroglyphics carved inside the Pyramid of Unas in Egypt, describe just such a journey, in which the pharaoh travels beyond the galaxy to be given knowledge, and granted immortality, by the gods.[167] This theme of the flight of the soul in search of some higher truth became an archetypal experience, enduring for thousands of years in almost every spiritual and philosophical tradition, until it eventually found an unlikely home in the mind of a modern physicist.

Not that Einstein's visual thought experiments were *identical* to visionary religious experiences, of course. Einstein made use of the same imaginative abilities that had been used by spiritual practitioners for millennia, but with his usual ingenuity he repurposed them for his own ends. Rather than trying to attain some clichéd religious experience that would confirm his preexisting beliefs, Einstein used his imagination to reach a new understanding of physical reality.

But he still saw these imaginal excursions as spiritual experiences, and the revelations he received still evoked a sort of religious ecstasy. Nowhere is this more obvious than in Einstein's greatest creation, the general theory of relativity. Soon after it was published, general relativity proved to be a triumph: its incredibly precise predictions were confirmed by the most meticulous experimental observations.[168] "Imagine my joy," Einstein wrote to a friend after the news was announced. "I was beside myself with ecstasy for days."[169] He'd experienced firsthand that to discover Nature's laws was to touch the divine.

He wasn't the first to feel he'd fathomed the mind of God with mathematics. All the scientists Einstein saw as saints of the cosmic religion had had similar experiences. When Kepler discovered the laws of planetary motion, he fell to his knees and cried, "My God, I am thinking Thy thoughts after Thee!"[170] The frontispiece of Newton's *Principia Mathematica* showed him sitting among the clouds being instructed, in person, by an angelic mathematical muse. And Galileo pointed out that Pythagoras "admired the human understanding and believed it to partake of divinity simply because it understood the nature of numbers," noting that he shared "the same opinion."[171]

Einstein's friend Bertrand Russell understood that this interplay of reason and revelation was inherent to all mathematical insight. "For all who were inspired by Pythagoras," Russell wrote, mathematics "retained an element of ecstatic revelation."[172] Russell realized this might sound strange to some, but "to those who have experienced the intoxicating delight of sudden understanding that mathematics gives, the Pythagorean view will seem completely natural."[173] Einstein agreed. "He who finds a thought that lets us penetrate even a little deeper into the eternal mystery of nature," he said, "has been granted great grace."[174] He even compared scientific work to religious practice. "The state of mind which enables a man to do work of this kind is akin to that of the religious worshiper," Einstein declared. "The daily effort comes from no deliberate intention or program, but straight from the heart."[175] And the scientific results themselves were almost like a religious revelation. "The genuine scientist is not moved by praise or blame, nor does he preach," Einstein said. "He unveils the universe and people come eagerly, without being pushed, to behold a new revelation: the order, the harmony, the magnificence of creation!"[176]

Yet Einstein never sought any supernatural explanation for the hidden harmony his own efforts helped make manifest. "I am not a mystic," he always insisted. "Trying to find out the laws of nature has nothing to do with mysticism." Nonetheless, reality still elicited his reverence. "In the face of creation I feel very humble," he admitted. "It is as if a spirit is manifest infinitely superior to man's spirit. Through my pursuit of science, I have known cosmic religious feelings. But I don't care to be called a mystic."[177]

Despite all his allusions to divine inspiration—and his deep respect for the revelatory nature of human ingenuity—Einstein was adamant that "experience is the alpha and the omega of all our knowledge of reality."[178] Truth was built from the bottom up, rather than transmitted from the top down. The real magic happened *within* the human mind. Einstein insisted that all of "our inner experiences," even the most ingenious insights, "consist of reproductions and combinations of sensory impressions."[179] For Einstein, "the function of pure reason" was to perform a kind of inner alchemy, transmuting the *prima materia* of raw perception into pristine mathematical models that matched the mysterious roots of reality with almost miraculous precision.[180] Somehow our tiny, localized minds derived laws of Nature that were applicable always and everywhere. And for the cosmic religion, these mathematical truths were the temples that would stand for all time, monuments raised to Reality by reverent minds. Or as Einstein expressed it: "Equations are for eternity."[181]

Not that mathematical intuition was infallible. Einstein readily acknowledged that "experience remains, of course, the sole criterion of the physical utility of a mathematical construction."[182] And yet mathematical models were far more than just a fabrication of the human mind. "Even if the axioms of the theory are proposed by man," Einstein said, "the success of such a project presupposes a high degree of ordering of the objective world."[183] Einstein insisted that the stupendous success of science over the centuries "justifies us in believing that nature is the realization of the simplest conceivable mathematical ideas." And therein, he thought, "lies the theorist's hope of grasping the real in all its depth."[184] This conviction that our cosmos is math made manifest is nothing other than the primary Pythagorean credo: *everything is number.*[185] Or as Einstein put it: "All nature expresses mathematical simplicity."[186] So "in a certain sense," he said, "I hold it true that pure thought can grasp reality, as the ancients dreamed."[187]

Modern physicists have pointed out that there's something mighty puzzling about this "unreasonable effectiveness of mathematics in the natural sciences."[188] But more than two millennia ago, the Pythagoreans were already marveling at the human mind's mysterious efficacy and wondering what it implied about our innermost essence. They came to an astonishing conclusion: our incredible capacity to comprehend the cosmos indicated a deep affinity between mental microcosm and material macrocosm. We were able to understand the overarching order of existence because that same order was the underlying essence of ourselves. For the Pythagoreans, the divine order was not only disseminated all around us—it was immanent everywhere, manifesting within our own bodies and minds at all times.

It probably didn't take Einstein long to discover that the Pythagoreans had held such heretical notions. In one of the first philosophical books he read as a young man, he almost certainly came across Schopenhauer's praise of "the wisdom and knowledge of Pythagoras" and his declaration that "the Pythagorean doctrine was a decided pantheism."[189] By no means was this Schopenhauer's own arbitrary conclusion. An ancient Pythagorean maxim implored us to "adore the immensity of God, who fills the universe."[190] According to the classical writer Clement of Alexandria, the Pythagoreans "say that God is One; and he is not, as some suspect, outside the universal order and separate, but within it...mind and living principle of the whole circle, movement of all things."[191] And even the early Church Fathers, who abhorred such heresy, agreed. Lactantius (c. 250–325 C.E.) described how the Pythagoreans saw the divine "diffused through all parts of the universe, and through all nature."[192] Contemporary scholars concur with the appraisals of the ancients: for the Pythagoreans, "the universe is divine."[193]

And yet only tiny pieces of the Pythagoreans' pantheist teaching survive today. Their most profound insights are preserved only in arcane proverbs and half-forgotten aphorisms. Such a fragmentary philosophy could never be enough to command Einstein's spiritual commitment. But in his extensive and eclectic reading, Einstein would soon discover a fully developed system of the immanent divine—rigorous, dispassionate, logical to a fault—in the work of one of the most singular sages in all of Western philosophy: Spinoza.

5

THE IMMANENT DIVINE

The God Spinoza revered is my God, too: I meet Him every day in the harmonious laws which govern the universe.

—Albert Einstein, in conversation with William Hermanns

MANY MYSTICS OVER THE MILLENNIA HAVE USED MARVELOUS META-phors to convey their direct personal experience that the divine is present everywhere. Countless sages across the ages have spoken of all seemingly separate things as ripples upon a single sea, sparks scattered from a single fire, delicate dewdrops destined to merge once more with the morning mists. But back in the seventeenth century, there was one weird mystic who tried to demonstrate the immanence of the divine not just with a rhetorical flourish, but in a rigorous fashion. He put forth his pantheism in a series of propositions and proofs—and what he hoped to prove was that mind and matter were merely two different manifestations of a single sacred Substance. It took a special kind of mind to deduce Nature's divinity, step by step, from a simple set of axioms. Never has pantheism been proclaimed in such a punctilious fashion. Never before, and never since, has the world seen such a meticulous mystic.

Most people find Baruch Spinoza's mathematical style exasperating, but Einstein must have thought manna had fallen from Heaven. For the man who'd experienced ecstasy proving the Pythagorean theorem as a child, whose favorite boyhood book was Euclid's geometry, and whose highest hope was to reveal the essence of reality with elegant equations, Spinoza's geometrical method was mesmerizing. In Spinoza, Einstein found not only his favorite philosopher, but an almost perfect prototype for his cosmic religion. Here was a message he could agree with, a method he could admire, and a mind he could adore. "For me, he is the ideal example of the cosmic man," Einstein said, and he declared Spinoza to be "one of the deepest and purest souls our Jewish people has produced."[1]

But in his search for the divine, Spinoza delved down far below the roots of the Jewish religion. Deep in the depths of his own soul, he discovered the wellspring of the world, the secret source that sustained all things. Spinoza basked in the waters of this sacred spring until they saturated his whole being. And if we wish to absorb the essence of Einstein's spirituality, we must dare to drink from the same eternal well that was the source of Spinoza's wisdom.

Spinoza was a pure pantheist—someone who believes the divine is everywhere, or every*thing*, or anyway something along these lines.[2] The poet Novalis praised him as a "God-intoxicated man," while the philosopher Hegel grumbled that "with him, there is too much God."[3] He was not the first pantheist in European history, but he was certainly the most important. And yet Spinoza was maligned in his own time for daring to say the world was divine, and he's mostly been neglected in the centuries since.

Baruch (or Benedict) Spinoza was born in 1632 in Amsterdam's Jewish ghetto, the descendant of Sephardic Jews who'd been hounded out of Spain and Portugal and wandered Europe before settling in the

Netherlands. He hardly knew his mother, Ana, who died when he was only six years old. His father, Michael, was a pious member of the local synagogue and a diligent practitioner of the merchant's trade. Most likely there was an expectation that Spinoza would follow in his father's footsteps, but the world of business would not be Baruch's path. After he entered the local Hebrew school around the age of seven, his intellectual gifts became apparent to all around him; soon he was the star student of his yeshiva (the old Hebrew word for a seminary school).

Spinoza seemed destined for the rabbinate until he diverged from Jewish doctrine in rather dramatic fashion. Although he studied the old authorities on Jewish theology with ardor, Spinoza was surrounded on all sides by scandalous new systems of thought.[4] "In every corner of Holland there was talk about the great questions and events of the epoch," as one biographer put it. "It was a political and rebellious century. And Baruch wished to be informed about it all."[5] Much like Pythagoras two millennia prior, Spinoza found himself born into a culture in full ferment. The list of thinkers living in his time reads like a roll call of history's greatest heretics. Giordano Bruno had been burned at the stake just three decades before Spinoza was born; Galileo's trial took place when Spinoza was a baby; and René Descartes was hiding out in Holland throughout Spinoza's boyhood.

Heresy was in the air—and, increasingly, also in print. Suppressing the speech of infidels had always been annoying for the ecclesiastical authorities, but banning their books was basically impossible. The printing press, which had made it so easy for Gutenberg to promulgate the Bible, also made it a simple matter to spread prohibited works. Needless to say, banned books weren't read in Spinoza's traditional Hebrew school. But when he was around twenty years old, he started studying with a teacher who was eager to introduce him to all the latest innovations in science, politics, philosophy, and mathematics: Franciscus van den Enden. Van den Enden (1602–1674) had originally been educated by the Jesuits but was expelled from the order for unclear reasons. He then moved around the Low Countries teaching Latin, Greek, and literature. Eventually he landed in Amsterdam and became a member of a group of radical thinkers known as the Amsterdam Circle. He tried earning a living as a bookseller, but when this failed he settled on starting a private Latin school.

But van den Enden's academy was much more than a mere language school. Latin was a gateway to lush new gardens of thought and feeling. The ancient language offered access to the once-lost wisdom of the classical past—Plato and Aristotle, Seneca and Cicero, Ovid and Epictetus—but it was also the lingua franca of living philosophers and scientists. And so Latin allowed van den Enden to flood his pupils' minds with all the new knowledge inundating Europe. The language of the ancients thereby led to Spinoza's first encounters with the empiricism of Francis Bacon, the physics of Galileo, the skepticism of Descartes, and the pantheism of Giordano Bruno. The overall impact on his young and impressionable mind is impossible to exaggerate. "All his life he had lived penned in the ghetto, his mind haltered by rabbinic thought," wrote one biographer. "And Van den Enden set him free."[6]

But just as Spinoza's mind was being opened to all these possibilities, other doors were closing forever. In quick succession, he lost almost all links to his old life. First came the death of his father in 1654, when Baruch was just twenty-two years old. His sister Rebeka, determined that nothing should go to her heathen brother, sought to keep the meager family estate for herself. And although Spinoza won the case in court, he contemptuously renounced his share of the inheritance, keeping only a bed and a blanket, and cut all connection with his family. He packed up his books, fled the family home, and went to live with his freethinking teacher. But van den Enden's school had gradually lost its good reputation and was now known as a nidus from which the infection of atheism was spreading. It had long been whispered that Spinoza, too, though he hid the symptoms well, was infected by the disease of disbelief. And now that he'd gone to live with an atheist agitator, the diagnosis could no longer be doubted.

Spinoza had been diverging from the orthodox path for some time, but eventually the Jewish authorities could no longer ignore his apostasy. The animosity Spinoza aroused among the orthodox is understandable if

we remember that the Jewish community in Amsterdam had fled ferocious persecution and forced conversion just a generation before Spinoza's
time. They were newcomers to the Netherlands—tolerated, yes, but by
no means assimilated, much less accepted as equals. Even in progressive
Holland, their political position was precarious. The last thing the Jews
of Amsterdam wanted was someone like Spinoza stirring up resentment
and causing yet another round of persecutions.

Still, the rabbis who served as the spiritual leaders of this tenuous
community were saddened to see so promising a pupil straying from the
path, and they made repeated bids for Spinoza to return to the fold. But
after a *niddui* (short ban) did not bring Spinoza to heel, they felt they had
no alternative but the *herem* (great ban). They declared that "we excom-
municate, expel, curse, and damn Baruch de Espinoza with the consent
of God," and Spinoza was cut off forever from the community of Israel.[7] The rabbis were reluctant to make him pay the ultimate price, but
the prevailing feeling among the mob was that Spinoza should be put to
death. After narrowly dodging the dagger of an assassin one night, Spinoza knew it was time to leave town. When he finally fled Amsterdam,
he was leaving behind all he'd ever known.[8]

And yet the end of his old life was the beginning of a new one. Spinoza was dead to his family, to his former friends, and to the Jewish
theologians; but now rebirth became a real possibility. "After experience
had taught me the hollowness and futility of everything that is ordi-
narily encountered in daily life," he later wrote, "I resolved at length to
enquire whether there existed a true good." What he wanted to know was
"whether, in fact, there was something whose discovery and acquisition
would afford me a continuous and supreme joy to all eternity."[9] Expulsion
from the Jewish community, the hatred of fanatics and fools, a life of
poverty and solitude—for Spinoza these were small prices to pay in his
search for the eternal.

So far as we can tell, he never once wavered. For the next two decades
he led a solitary life, renting a series of small cottages in Rijnsburg, Voor-
burg, and the Hague. His was a simple existence dedicated to scholarly study and the systematic exposition of his soon-to-be-infamous

philosophy. Despite all the accusations of his infernal atheism, by all accounts he led an almost saintly life. Contemporaries noticed in him "no lack of morality or of exemplary conduct," reporting that "he spent his life in peace and in celibacy."[10] One early biographer recalled how "he knew admirably well how to be the Master of his Passions," and another asserted that "one always found him in an even and agreeable humor."[11] Even his enemies, although appalled by his beliefs, admired his actions. They cursed him for his blasphemous convictions but could find no fault in his blameless conduct.

Along with his infamy among the ignorant, Spinoza soon acquired a well-deserved fame among thinkers and philosophers. Although he was a poor man, without academic appointment or political position, his writings were revered for their daring and clarity, and his handwritten manuscripts were circulated in secret.[12] It wasn't long before his books made it into the right hands—and his ideas into the right minds. A small but steady stream of intellectual luminaries began to correspond with him through letters; some even came to visit the curious sage in his small cottage. Among the illustrious admirers were Gottfried Wilhelm Leibniz, philosopher and co-inventor of calculus, and Christiaan Huygens, one of the founders of modern physics and a living legend even in his own day.[13]

Spinoza had such stature among the intellectual elite that, in spite of the scorn of the theologians, he was offered a cushy academic position at the University of Heidelberg. Under the auspices of the progressive prince Karl Ludwig, he was promised "the most extensive freedom in philosophizing" and "the pleasure of living a life worthy of a philosopher."[14] But after his debacle with the Jewish community, Spinoza feared that the position would impose upon his intellectual freedom, and that public lecturing would arouse even greater animosity among the ignorant.[15] Determined to protect both his independence and his peace of mind, he declined.

Instead, Spinoza supported himself through a humble trade, and he could hardly have picked a more perfect profession. He chose to work as a lens grinder, fashioning glass lenses that found their way into the first formidable instruments of the Scientific Revolution: microscopes and telescopes. Spinoza was no amateur. Astronomers and anatomists, physicians and physicists all praised his lenses for their precision and put them to use in the pursuit of new knowledge. Huygens proclaimed his lenses "very excellent," and Leibniz called him "an outstanding optician," writing directly to Spinoza to rave about his "remarkable skill in optics."[16] An old colleague of Spinoza's who'd gone on to become a doctor noted that "I own a first-class microscope made by that Benedictus Spinoza, that noble mathematician and philosopher, which enables me to see the lymphatic vascular bundles."[17]

It's hard to think of a more metaphorically perfect métier. At night, Spinoza labored to produce a profound philosophy that would point the way to a more perfect truth; by day, he ground glasses that gave the first modern scientists the means to gaze up at the grand spectacles transpiring among the stars and to look into the little marvels within living beings. With every waking hour he worked to fashion instruments, be they physical or philosophical, that would expand the horizons of the human mind.

By making his own living, Spinoza was free to make up his own mind; and by making the most of his remarkable mind, he made history. The first polemic to come from his pen was his *Theological-Political Treatise*, and the spirit of scientific skepticism is palpable on almost every page. According to one historian, the furor created by its publication in 1670 was, "without question, one of the most significant events in European intellectual history."[18] Europe was stunned by Spinoza's short screed, in which he dismissed prophecy and miracles, denied the immortality of the soul, demanded the separation of church and state, and demonstrated that the Bible was no revelation but rather a clumsy compilation of diverse writings—"merely notes and collations," as he put it.[19] Spinoza's "godless document" was famously branded "a book forged in hell" and officially banned. And although the author of this assault on orthodoxy was anonymous, word got around. Soon Spinoza was infamous for his alleged atheism.[20]

But as Einstein once said, "mere unbelief in a personal God is no philosophy at all," and Spinoza likewise had no intention of stopping at skepticism.[21] Although he was denounced by the orthodox as "an atheist, a scoffer at religion," Spinoza's dream was not to denigrate the divine but rather to demonstrate that it was disseminated everywhere.[22] And in his subsequent works, the immanent divine became Spinoza's central theme. He argued that a single inscrutable Substance was the substrate of all things—everything around us and everything within, matter and mind alike.[23] For Spinoza, this Substance was "conceived through itself" and consisted of "infinite attributes," all of which were simply expressions of an "eternal and infinite essence."[24] We could call it whatever we wanted—Substance, Nature, or even God—but as far as Spinoza was concerned, "it is the same, or not very different, to assert that all things emanate necessarily from God's nature and that the universe is God."[25] From this seemingly simple assertion, he concluded that "all things are united through Nature, and they are united into one, namely, God."[26]

Spinoza's contemporaries were convinced that this made him an atheist, and it's a common misconception even today that Spinoza's God is identical with the universe we see around us—no more and no less than a bunch of atoms whizzing around in the void. In the words of the illustrious atheist Richard Dawkins, pantheism is nothing but "sexed-up atheism."[27] But "the supposition of some," Spinoza said, "that I endeavour to prove...the unity of God and Nature (meaning by the latter a certain mass or corporeal matter) is wholly erroneous."[28] Spinoza's God wasn't just a bunch of static *stuff* sitting around in space. In Spinoza's lingo, matter was merely one *mode* in which the infinite Substance manifested itself. This divine Substance was more than just atoms and energy; it was diverse and dynamic, bountiful enough to create both conscious beings and an entire beautiful cosmos. And all the while, it remained intrinsic in everything, "the immanent cause," as Spinoza insisted, "of all things."[29]

For Spinoza, all things meant *all* things: not just matter, but also mind. "The Eternal Wisdom of God," he said, "has manifested itself in all things and especially in the human mind."[30] In words Einstein would echo centuries later with his assertion that he was a part of Infinity, Spinoza saw every human being as "a part of the whole of Nature."[31] And the highest goal of his religion was to realize this fact. "The supreme good," Spinoza said, was to arrive at "the knowledge of the union which the mind has with the whole of Nature."[32]

With the benefit of hindsight, it's easy to see Spinoza's pantheism as a mirror image of many Eastern teachings. Almost every serious student of Spinoza's philosophy has noticed the striking similarities, and for centuries he's been seen as a sort of spiritual prodigy whose true home was the East. No less a luminary than Immanuel Kant spoke of Spinoza's philosophy as equivalent to the pantheism of Chinese Taoism and Tibetan Buddhism.[33] Schopenhauer similarly argued that India was Spinoza's "true spiritual home."[34] And even Einstein spoke of Spinoza and Buddha in the same breath as "enlightened," acknowledging that "most philosophers are indebted to the Hindus [but] Spinoza's contribution springs from his own brain."[35] Yet for all his affinities with Eastern religion, the parallels between his pantheism and certain Eastern systems are completely coincidental. He must never have known that his philosophy so closely resembled the religious teachings of another age and a distant land.[36]

Although Spinoza was unaware of Eastern spirituality, he was an expert in Western theology. He had a scholar's grasp of Judaism and Christianity—including being able to read the Bible in the original Hebrew and Greek. But for all his intimacy with mainstream Western religions, these are also unlikely to have inspired his radical pantheism. Although Spinoza was a diligent student for many years at the yeshiva, his pantheist notions were anathema to Jewish orthodoxy. Judaism saw the cosmos as God's created artifact; Spinoza saw it as its own autonomous Artificer, "a Being existing through its own sufficiency or force."[37] Equating God with the world was unthinkable within the confines of any ordinary Jewish theology.

Pantheism might have been a bit more acceptable among adherents of kabbalah (Jewish mysticism), who sought to understand the essence of *ein sof* (the Infinite) through mystical experience. The Jews who'd been expelled from Spain and Portugal took Jewish mysticism along with them, and both kabbalistic beliefs and books were commonplace in the Dutch exile community Spinoza had once called home.[38] This has led some scholars to suggest that kabbalah might have been the source of his pantheism.[39] And Spinoza does tell us that he read the works of "a number of Cabbalistic triflers." But far from finding any inspiration in their writings, he was appalled by their irrational mysticism, complaining that "their madness passes the bounds of my understanding."[40]

Still, Spinoza's ideas didn't just spring up out of nowhere. And there's one very plausible person who probably inspired his pantheism. Giordano Bruno (1548–1600) lived just a generation before Spinoza and, as everyone knows, was burned at the stake by the Inquisition for his heretical ideas. Bruno was by far the most prominent person painting a pantheistic portrait of the world in Spinoza's time, and although his books were banned (and, like their author, burned), they still circulated in secret among the freethinkers of the seventeenth century. Most scholars think Bruno's influence on Spinoza is likely, and many see it as self-evident.[41] Even those who are skeptical of any direct influence agree that "certainly there are striking similarities," and admit that "the central purpose and conception of their philosophies are the same"—namely, to show that "all things flow from and are controlled in a unified reality which is self-contained and self-sufficient."[42]

Historians might never come to complete agreement on Spinoza's sources, but Bruno's intellectual origins aren't hard to trace. In his radical pantheist work *Cause, Principle, and Unity* (which many feel certain Spinoza read), our old friends the Pythagoreans make an appearance on practically the first page.[43] "We may demonstrate with Pythagoras," Bruno wrote, "how an immense spirit, under different relations and according to different degrees, fills and contains the whole."[44] Like Pythagoras before him, Bruno saw the divine dancing in even the most menial of objects and considered "matter to be an absolutely excellent and divine thing."[45] So perhaps Spinoza was simply carrying on the old Pythagorean tradition that all was One.

But assuming Giordano Bruno and his Pythagorean precursors were a major inspiration for Spinoza's pantheist message, there's still the mystery of his peculiar *method*. Rather than trying to persuade with reasoning (the typical tactic of philosophy) or inspire with imagery (the usual strategy of spiritual teachers), Spinoza presented his pantheism as if he were elucidating a mathematical proof. In his magnum opus, the *Ethics*, he made a self-conscious effort to model his mode of expression on the famous mathematician Euclid (the book's full title is actually *Ethics, Demonstrated in Geometrical Order*). In a work almost unprecedented in prior millennia and unparalleled in the centuries since, Spinoza tried to demonstrate his daring idea of the immanent divine in the driest style imaginable. Axioms and propositions are followed by proofs, definitions are followed by deductions, and the argument is quickly complicated by complex corollaries and self-referential scholia.

It's a strange choice (to say the least), and along with Spinoza's baroque terminology it makes for something almost unreadable. But Spinoza had his reasons for adopting this soporific style. According to one modern expert, the "choice of the axiomatic method represents nothing more, and nothing less, than an awesome commitment to intellectual honesty and clarity."[46] Another eminent Spinoza scholar agrees that "the axiomatic style mirrors the system's rationality and exemplifies the way knowledge should be grasped."[47] Then as now, the meticulous method of mathematical proof was regarded as the epitome of clear communication between honest minds with nothing to hide. This was why Spinoza sought to share his highest mystical experiences with mathematical exactitude. "He wanted to build a temple of reason," as one biographer put it, "in which each stone stood square and firm on the one beneath it."[48]

Einstein agreed that the clear exposition of a complex truth was the aim of the *Ethics*, but he admitted that Spinoza's style made him almost incomprehensible. "Spinoza is, among the great classical thinkers, one of the least accessible," Einstein realized, "because of his rigid adherence to the geometric form of argumentation." This convoluted method caused considerable confusion. "Spinoza thereby made it difficult for the reader," Einstein recognized, "who all too quickly loses patience and breath before he reaches the heart of the philosopher's ideas." And to this day, Spinoza's deep and delightful philosophy is reserved for the determined few. All the same, Einstein saw that his abstruse approach had its advantages. "Throughout Spinoza's writings," Einstein wrote, "one will find sharp and clear propositions which are masterpieces of concise formulation."[49]

Euclid probably would have been proud. But it's a little poignant that the main reason almost no one reads Spinoza today is the same reason Einstein admired him so much: he presented his unifying mystical pantheism with Euclid's mathematical precision.

Euclid himself is an enigmatic figure. As with so many ancient thinkers, almost nothing is known of the man. But careful sifting of the scant evidence has led historians to conclude that "Euclid received his mathematical training in Athens from the pupils of Plato."[50] Plato, for his part, had learned his geometry and mathematics from the Pythagoreans of Magna Graecia, whom he went to live with and learn from after the execution of Socrates.[51] And despite the hazy details of Euclid's personal life, it's obvious from a close analysis of the *Elements* itself that much of the material was borrowed from Pythagorean sources.[52] Although these links are little known and even less appreciated today, Einstein almost certainly knew that a continuous tradition connected Euclid to the Pythagoreans.[53] So once again we see the sages of southern Italy exerting a strong but subtle influence on the meandering course of the human mind. Almost invisible on the surface of intellectual history, a secret stream of mathematical mysticism ran underground for centuries, flowing from Pythagoras to Plato, Plato to Euclid, Euclid to Spinoza— and finally from Baruch Spinoza to Albert Einstein.

Einstein venerated Spinoza's pantheist vision, and he couldn't keep his heretical convictions quiet forever. After worldwide fame came to him in 1919, it was only a matter of time before his personal sympathy for pantheism would suffer from public scrutiny. By 1929, many average people were eager to comprehend Einstein's spiritual views, but many religious authorities were just as eager to condemn his heretical beliefs. The accusation that he was an atheist was widespread, including a denunciation from the cardinal of Boston prominently published in the *New York Times*.[54] Alarmed at the cardinal's outrage, a rabbi in New York sent Einstein a telegram that read simply, "Do you believe in God?" Einstein was astounded by the audacity of the request, but he replied all the same.

Some of Einstein's books on Baruch Spinoza at the Albert Einstein Archives in Jerusalem. Spinoza believed that "the Eternal Wisdom of God has manifested itself in all things and especially in the human mind." And much like Einstein three centuries later, Spinoza said that "the supreme good" was to arrive at "the knowledge of the union which the mind has with the whole of Nature." These heretical ideas led to him being branded "an atheist, a scoffer at religion," but Einstein venerated Spinoza's pantheist vision. "For me, he is the ideal example of the cosmic man," Einstein said, and he declared Spinoza to be "one of the deepest and purest souls our Jewish people has produced." When someone once asked Einstein if he believed in God, he replied, "I believe in Spinoza's God, who reveals himself in the lawful harmony of the world." Einstein had such high esteem for Spinoza's magnum opus, the *Ethics* (Einstein's German edition is pictured above), that he even insisted that "there is nothing more ardently to be wished for than that this book may influence an ever-wider circle of human beings and thus exert its spiritual force on individuals and society alike." Photo © the author (2023), used with permission of the Albert Einstein Archives.

"I believe in Spinoza's God," he wrote, "who reveals himself in the lawful harmony of the world, not in a God who concerns himself with the fate and the doings of mankind."[55]

Although Einstein himself confessed that "I do not know if I can define myself as a Pantheist," he did admit to feeling "fascinated by Spinoza's Pantheism" and elsewhere gushed about his "great admiration for the man."[56] In fact, Einstein adored Spinoza's *Ethics* so much that one of his deepest desires was that it be disseminated to as many people as possible. "There is nothing more ardently to be wished for," he felt, "than that this book may influence an ever-wider circle of human beings and thus exert its spiritual force on individuals and society alike."[57]

And although we'll probably never know for sure whether Spinoza was familiar with Giordano Bruno's pantheist books, we do know that Einstein definitely was. He was probably first exposed to the basics of Bruno's worldview through his thorough reading of Schopenhauer. Schopenhauer's erudition was immense, and in elucidating his own views he also managed to present a masterly synthesis of twenty-five centuries of world philosophy, East and West. He believed Bruno to be "a great man" and placed him squarely in the pantheist tradition alongside Pythagoras, Spinoza, and the authors of the Upanishads.[58] He even suggested that both Bruno and Spinoza "would have led a peaceful and honoured life among men of like mind" if only they'd lived along "the banks of the sacred Ganges" in ancient India, which Schopenhauer considered "their true spiritual home."[59] Throughout his philosophical writings, Schopenhauer also touched on Bruno's core beliefs: namely, that a single transcendent Substance underlies all things, and that death is an illusion to the sage who has seen that our essence transcends time and change.[60]

In addition to this secondhand exposure through Schopenhauer, it also seems that Einstein went straight to the source, as he so often did, and read Bruno's own words. Among the books of his personal library there's a little anthology of Bruno's writings, including selections from the pantheist dialogues and cosmological speculations that got him into so much trouble with the Inquisition. From a handwritten inscription on the book's first page, we can see that someone (they didn't sign their name) gave the book to Einstein in 1929, and apparently they felt the great physicist warranted comparison with the famous heretic. "Giordano Bruno: This you had been, this you are," they wrote.[61]

Einstein seemed to share the sentiment: although there's no way to know how much of this little book he read, if any (there are no handwritten markings within), he obviously knew enough about Bruno to admire his philosophy and empathize with his fate. "Maybe some centuries ago I would have been burned or hanged," he once mused. "Nonetheless, I would have been in good company. Giordano Bruno and the heretics through the ages were often people with deep religious feelings."[62]

Born in Italy and trained as a Dominican friar, Bruno was excommunicated from the Catholic Church for reading banned books and endorsing banned beliefs. After fleeing the Inquisition in Italy, he spent some two decades wandering Europe and writing down his wild ideas about an infinite cosmos that contained countless other worlds like our own—worlds on which, undoubtedly, beings like ourselves contemplated their own existence and worshipped their own gods. Along the way, he lectured at many of the great European centers of learning still renowned today (in Oxford, Paris, Geneva, and many others) and established an international reputation as a peculiar kind of polymath. Part scientist and part spiritual seer, he became famous as a mystic and magician, a mathematician and expert in the art of memorization.[63]

But at his core, Bruno was a pantheist par excellence. Apart from Spinoza, perhaps no other mind in the early modern era was so utterly intoxicated with the divine or so obviously in awe of Infinity. For Bruno, all things were "parts of the Infinite," and in one of the poems included in his scientific dialogues about the cosmos, he tells us of his "solitary passage to those parts [the heavens], born of my thoughts, rising to the courts of infinity."[64] He leaves little doubt that he's referring to a flight of the soul, to his own inner experience of the Infinite. "I affix these wings, leap in the air, and plow the skies, 'til in the infinite I stand," he writes.[65] But Bruno didn't present his ideas as religious dogma requiring blind faith, or direct revelation dependent on his own authority. Instead, much like Spinoza just a generation later, Bruno opted to explain his "contemplation of the infinite" as clearly as he could: "through demonstrative arguments," he explains, "arriving like true ambassadors of objective Nature, presenting themselves to the searcher, appearing to the observer, clear to those who would understand, plain to those who would comprehend."[66]

And for Bruno, "the splendor of the divine" offered more than just an uplifting experience.[67] An intimacy with Infinity conferred a kind of philosophical freedom—a sort of salvation, even. "Only the singular one is my beloved," he writes in an ode to the Infinite. "Through her I have freedom in subjection, happiness in sorrow, wealth in poverty, and life in death."[68] And with heartfelt gratitude, he thanks the Infinite for the gifts it has given him: "I thank you, my sun, my divine light....You have led me to an exalted place, your attentions have healed me."[69] By no means did Bruno think of his encounter with Infinity as his own special endowment, inaccessible to the rest of us. On the contrary, he hoped his writings could convey the immense value of this vision and even elicit an equivalent experience in others. May "the spirits be awoken," he prayed, "and the hearts be opened of all who suffer in darkness."[70] No wonder Einstein admired him.

And no wonder the Inquisition was appalled. For centuries, the Catholic Church had seen itself as the necessary instrument and intermediary of our salvation: sole possessor of the actual truth and the only means of access to the afterlife. Then along came Bruno, claiming that the Christian religion was not only unexceptional, but unnecessary—that God was immanent everywhere and accessible to everyone. Historians have long argued over why Bruno incurred the wrath of the Inquisition. Was he the first martyr to free thought and scientific inquiry, or was he just another religious zealot who paid the ultimate price for antagonizing the powers that be? There is truth to both views, because there wasn't any *one* belief that got Bruno burned alive. His indictment included no less than eight counts of heresy, and the Inquisition took issue with both his scientific speculations and his spiritual convictions.[71] But there's no doubt that his interest in the Infinite was an important issue, and at least one biographer believes that the "most terrible" of Bruno's heresies was his insistence "that we live in an infinite universe and that innumerable worlds exist upon which creatures like ourselves might thrive and worship their own gods."[72]

Whatever the reasons why, his fate was sealed. After eight years of imprisonment and torture, Bruno was condemned to death as a heretic by the Pope himself. His tongue was placed in a wooden vise to silence his wicked words; he was hung naked and upside down to humble his pride; and finally, he was burned alive to annihilate his body.[73] Arthur Schopenhauer, for one, was sickened by the actions of these "fiendish fanatics," and asked us to "imagine the tender, spiritual, thoughtful being in the hands of coarse and enraged priests as his judges and executioners."[74] Such was the cost of cosmic religious feelings in an uncomprehending era.[75]

Fortunately for Einstein, Bruno's grisly fate was not for him. But much like Bruno and Spinoza, Einstein was still seen as a heretic by the faithful, and he still paid a considerable price for his cosmic religious feelings. After introducing the public to his cosmic religion in 1930 in a special piece for the *New York Times Magazine*, he was savagely attacked. One Catholic bishop, Dr. Fulton Sheen, felt that "the wise and distinguished *New York Times*" had "degraded itself" by publishing Einstein's views. Without even an ounce of irony, Sheen accused Einstein of "being very dogmatic when he represents his cosmical religion without dogmas or church." To Sheen, the idea of a third phase of spirituality that transcended traditional teachings was totally impractical, almost tantamount to madness. "When he says that we have passed the stage of a religion of fear and morals," Sheen fumed, "he is talking the sheerest kind of stupidity and nonsense." Einstein's reverence for the real, concrete cosmos we happened to inhabit simply could not be taken seriously. "He is asking us to accept something that we never can love," Sheen lamented. "The test of love is the willingness to fight for a thing. Men are willing to die for what we call 'the milk of human kindness,' but who in this world is willing to lay down his life for the Milky Way?" The unspoken assumption here is that one should be willing to die for religious dogma (and presumably kill for it, too, as Bruno's death so gruesomely demonstrates). Einstein, of course, was offering the exact opposite: a set of luminous ideals worth *living* for. But Sheen found it all laughable. "There is only one fault with his cosmical religion," Sheen concluded. "He put an extra letter in the word—the letter 's.'"[76]

Einstein was unfazed. Even his scientific ideas had elicited intense animosity from unscrupulous critics, so it was no surprise to him when his spiritual ideals also came under assault.[77] Never one to keep quiet when it came to the things that mattered most, Einstein kept on praising Spinoza's spirituality and promoting his own pantheist-Pythagorean world-view in public speeches and personal correspondence.[78] And throughout his long life, Einstein's enormous esteem for Spinoza never waned.[79] Ignoring the criticisms of his cosmic religion, he remained convinced that true science and true spirituality were completely compatible.

These convictions were put to the test a decade later, when Einstein agreed to contribute to a conference on science and religion in New York. More than five hundred audience members listened with mounting horror as his speech was read. Einstein argued that "human fantasy" had "created gods in man's own image." He maintained that "the main source of the present-day conflicts between the spheres of religion and of science lies in the concept of a personal God." And he held up Spinoza and Buddha as "religiously enlightened" exemplars of a more appealing alternative to the monotheistic orthodoxies. He concluded by telling a room full of rabbis, priests, and other religious believers that "in their struggle for the ethical good, teachers of religion must have the stature to give up the doctrine of a personal God, that is, give up that source of fear and hope which in the past placed such vast power in the hands of priests."[80]

The responses, ranging from the petulant to the preposterous, were predictable. "There is no other God but a personal God," proclaimed one Catholic priest. "Einstein does not know what he is talking about. He is all wrong."[81] *Time* magazine, which would later lionize Einstein as the "Person of the Century," whined that "Einstein's message was the only false note of the entire conference."[82] And a historian harangued Einstein for thinking that science could complement spirituality: "God is a spirit and cannot be found through the telescope or microscope," he wrote. "As everyone knows, religion is based on Faith, not knowledge."[83]

Einstein had nothing but contempt for his critics. "I was barked at by numerous dogs," he wrote, "who are earning their food guarding ignorance and superstition for the benefit of those who profit from it."[84]

Naturally, not everyone had a negative reaction to his cosmic religion. "The great leaders, thinkers and patriots of the past who fought and died for free thought, free speech, free press, and intellectual liberty arise to salute you!" wrote one veteran of the First World War. "With the great and mighty Spinoza, your name will live as long as humanity."[85] There was even a Christian theologian who declared that he "discerned nothing dangerous in the Einstein point of view." Einstein's idea of an impersonal, immanent divine was "an old concept...as old, say, as the Hindu religion."[86]

No doubt Einstein found these comparisons flattering. Like so many other pantheists through the ages, he refused to worship the otherworldly fantasies put forth by the orthodox religions, and instead insisted that *this* reality was worthy of our reverence. In true pantheist fashion, he proudly proclaimed that "nature is a perfect structure," and he made it plain that he had "great respect for matter."[87] This idea that everything was divine was not some extraneous aspect of Einstein's spirituality, but its very essence. "If something is in me that can be called religious," he insisted, "then it is the unbounded admiration for the structure of the world so far as science can reveal it."[88]

Einstein's reverence for reality came from the realization that the material world wasn't merely mindless mechanism. For Einstein, Nature was anything but inert: it was active and animated, imbued with an exuberant energy. "The whole of nature is life," as Einstein explained. "Every cell has life. Matter, too, has life; it is energy solidified."[89] Individual objects came and went, of course, but the substrate itself endured. "There is no permanence in matter," Einstein admitted, "but there is in energy."[90] And so he insisted that "energy is the basic force in creation," the everlasting "substance of the universe."[91] It's no accident that Einstein employed the strange term *substance*: it's the same word Spinoza selected for *his* ultimate substrate, which he considered synonymous with both God and Nature. And Einstein was equally explicit that his enduring substance was something sacred. "Whether mass is transformed to atoms, electrons, or motion, it's still a reality, a manifestation of eternal energy," he explained. "This oneness of creation, to my sense, is God."[92]

Despite his devotion to Spinoza, Einstein's brand of pantheism neither began nor ended with the *Ethics*. Actually, Einstein was exposed early on to a plethora of pantheistic systems, Eastern and Western. In his early twenties, he read Schopenhauer's acclaimed collection of essays, *Parerga and Paralipomena*, which included an encyclopedic synopsis of past thinkers.[93] Indeed, in the very first essay Schopenhauer proclaimed the pantheism of Bruno and Spinoza analogous to the teachings of the Vedas and the Upanishads, the oldest sacred writings of India.[94] So by the time Einstein encountered the *Ethics*, he almost certainly saw Spinoza's pantheism as a mirror image of ancient Indian monism.

Instead of the wondrous One of the Pythagoreans, or the sacred Substance of Spinoza, the ancient Indian sages believed in an omnipresent absolute they called Brahman. Brahman was a unity underlying all multiplicity, an "eternal infinite divine power" that "presents itself to us materialised in all existing things, which creates, sustains, preserves, and receives back into itself again all worlds."[95]

Although Einstein never mentions this old Indian term for absolute reality, it's impossible he was unaware of it. Schopenhauer, whose books Einstein read repeatedly, constantly quoted from the sacred scriptures of ancient India.[96] He venerated the Vedas as "the fruit of the highest human knowledge and wisdom," and he gushed that the Upanishads were "the greatest gift to the nineteenth century," because they "teach man to regard himself as Brahman, as the original being himself, to whom all arising and passing away are essentially foreign."[97] Understanding that our essence was eternal allowed us "to look death calmly in the face," and for Schopenhauer, *this* was the ultimate aim of philosophy: not rational understanding or logical consistency, but the attainment of "that peace that is higher than all reason, that ocean-like calmness of spirit, that deep tranquility, that unshakable confidence and serenity."[98] And he made it clear that he considered this kind of philosophical enlightenment as analogous to "reabsorption in Brahman."[99]

Einstein was also given glimpses of ancient Indian pantheism by Mahatma Gandhi. Although he never got to meet Gandhi in person, Einstein esteemed him above any other living human being, and he owned many books by the Mahatma, including his autobiography (*My Experiments with Truth*) and an expertly curated collection of his deepest thoughts on diverse subjects (*The Wit and Wisdom of Gandhi*). Einstein claimed to have studied the Mahatma's writings "with real admiration," and he no doubt noticed that Gandhi proclaimed a very pantheistic notion of the natural world.[100] "There is an indefinable mysterious power that pervades everything," Gandhi claimed.[101] "Whilst everything around me is ever changing, ever dying, there is underlying all that change a living power that is changeless, that holds all together, that creates, dissolves and recreates," he explained. "That informing power or spirit is God."[102]

Like so many of the pantheist thinkers Einstein admired, Gandhi insisted on this power's immanence. "God is not a Power residing in the clouds. God is an unseen Power residing within us and nearer to us than finger-nails to the flesh," he thought. "He is in every atom about us, around us and within us."[103] According to the Jesuit scholar J. T. F. Jordens, who scoured all ninety volumes of Gandhi's *Collected Works* to get at the essence of his spirituality, there's no doubt about where Gandhi got these ideas. "His concept of the absolute was essentially the Advaitic one of absolute non-dualism," Jordens concluded categorically. "It was *Brahman*."[104] And Gandhi in fact openly acknowledged that "I am myself a follower of the Advaita doctrine."[105]*

But Einstein probably didn't settle for Brahman secondhand from Schopenhauer and Gandhi. In his personal library lies a book, *The World's Great Scriptures*, that provides a sweeping survey of world spirituality, compiling many key texts in English translation—including substantial selections from the Upanishads.[106] We can't be sure Einstein actually

---

* *Advaita* literally means "non-dual." Advaita Vedanta is one the most famous philosophical schools of India.

111

read it, but if he did peruse its pages he would have found that ancient India's poet-mystics insisted that the divine was distributed everywhere in existence. "All this is Brahman," the Chandogya Upanishad asserts of the entire universe. "Let a man meditate on that visible world as beginning, ending and breathing in it."[107] Likewise, the Mundaka Upanishad explains that "Brahman shines forth grand, divine, inconceivable" from every part of the world.[108]

Whether or not Einstein actually read the ancient texts of the Upanishads, we know for a fact that he met multiple times with a modern mystic who embodied Eastern pantheism: Rabindranath Tagore (1861–1941). Although mostly forgotten today, in his time Tagore was almost as acclaimed in the arts as Einstein was in the sciences. A prolific poet, painter, playwright, and musician, he hobnobbed with legendary poets like W. B. Yeats and Ezra Pound, received a knighthood from King George V, and won the Nobel Prize for Literature (the first non-European ever to be accorded the honor). Besides being a highly versatile artistic virtuoso, Tagore was also very much a mystic and philosopher. His poetry is filled with allusions to his own spiritual experiences; his father was considered a maharishi (a great seer) and even founded a new religion, Brahmoism, which borrowed extensively from the Upanishads; and together, father and son founded an ashram in West Bengal that is active to this day.

In 1930, Tagore was invited to deliver the distinguished Hibbert Lectures at Oxford, and he used this prestigious public venue to propound his own personal brand of pantheism. He called it "the Religion of Man," and his vision was very much inspired by the nondualism of the Upanishads. In Tagore's system, individual existence was only a limited expression of something impersonal and infinite; consciousness was nothing other than the cosmos contemplating itself; and the overarching goal for the seeker was to cultivate a "consciousness of the infinite."[109] Although this Infinite was present in everything, it was also in some sense utterly beyond us, unknown and unknowable. Still, Tagore believed something *could* be said about the ultimate substrate of all things. "The positive aspect of the infinite," he explained, "is in *advaitam* [nonduality], in an absolute unity...an intense quality of harmony."[110]

A few weeks after giving his lectures at Oxford, Tagore visited Einstein in Germany. "We talked long and earnestly about my 'religion of man,'" Tagore later told an interviewer, "and by his questions I could measure the trend of his own thinking."[111] The two great minds didn't waste any time: they dived straight into a discussion of the divine.[112] "Do you believe in the Divine as isolated from the world?" Einstein asked. "Not isolated," Tagore replied. "The infinite personality of Man comprehends the Universe."[113] Tagore was talking about the old Upanishadic axiom that Atman is identical to Brahman. "According to Indian philosophy," he told Einstein, "there is Brahman, the absolute Truth, which cannot be conceived by the isolation of the individual mind or described by words, but can only be realized by completely merging the individual in its infinity."[114]*

Einstein's encounter with Tagore obviously made an impression. He packed his personal library with books of Tagore's mystical poetry, and not long after their meeting, Einstein authored an homage honoring Tagore's contributions to the human community.[115] "You have served mankind by a long, fruitful life," Einstein wrote, "spreading a mild spirit, as has been proclaimed by the wise men of your people."[116]

Just a couple months after he talked to Tagore, Einstein composed his first cohesive account of the cosmic religion, prominently published in the *New York Times Magazine*. And it can't be a complete coincidence that his third phase so closely resembled the religion of the rishis. Just as the ancient Indian sages had insisted that Brahman was utterly impersonal and devoid of human qualities, Einstein claimed that anyone who'd experienced the "highest kind of religious feeling" could accept "no God conceived in man's image."[117] Just as Advaita insisted that there was no actual difference between the external world and our inner nature (Brahman and Atman), Einstein wanted us to see that the same

---

* Incredibly, this specific sentence where Tagore tells Einstein about Brahman and the Indian ideal of the individual merging with the Infinite has been edited out of an otherwise complete transcript of their conversation provided in an Einstein biography. It seems like yet another example of Einstein scholars' tendency to downplay his awareness of, and interest in, Eastern spirituality. For the omission, see Pais, *Subtle Is the Lord*, 103.

Einstein with the Indian mystical poet Rabindranath Tagore (1861–1941), recipient of the Nobel Prize in Literature. Tagore and Einstein met in person three times to talk philosophy and spirituality. They first met at Einstein's summer home in Caputh, Germany, in 1930 (pictured here). Tagore told Einstein about the ancient nondualistic doctrine of the Upanishads that equates Atman and Brahman. "There is Brahman, the absolute Truth," Tagore explained, "which cannot be conceived by the isolation of the individual mind or described by words, but can only be realized by completely merging the individual in its infinity." Tagore felt that Einstein had experienced a similar sort of ego dissolution through science rather than spiritual practice, insisting that Einstein's "transcendental materialism" transported him to "the frontiers of metaphysics, where there can be utter detachment from the entangling world of self." Their meetings must have made an impression, because Einstein packed his personal library with books of Tagore's philosophy and mystical poetry. Photo circa 1930; public domain.

"sublimity and marvelous order...reveal themselves both in nature and in the world of thought."[118] And just as the Indian ideal was to liberate the individual from the ego and instead identify with the Infinite, Einstein likewise argued that, for the cosmic human being, "individual existence impresses him as a sort of prison and he wants to experience the universe as a single significant whole."[119] It seems obvious that, in elucidating his cosmic religion, Einstein everywhere echoed the ideas of ancient India.

But it wasn't just the Indian traditions that intrigued Einstein. It turns out he was also fascinated by ancient Chinese philosophy, and he wasn't the only great physicist who felt its allure. Erwin Schrödinger owned two copies of the Tao Te Ching.[120] Niels Bohr, Einstein's close friend and constant intellectual sparring partner, thought that quantum physics, with its emphasis on complementary opposites, was grappling with the same kind of conundrums that had confronted Lao Tzu, the Tao Te Ching's legendary author.[121] And Wolfgang Pauli often discussed Taoism and openly acknowledged that "when it comes to religion and philosophy, my background is Lao-tse and Schopenhauer."[122]

So it shouldn't astonish us that Einstein owned several works by those he admired as "the old Chinese sages," including the *Analects* of Confucius, the spiritual musings of the Taoist sage Chuang Tzu, and not one but *five* different copies of the Tao Te Ching.[123] This included the first English translation by a native Chinese-speaker, Sum Nung Au-Young, who penned a handwritten dedication on the first page: "To Dr. Einstein, with the best wishes and the highest esteem."[124] Also notable was an early German translation of the Tao Te Ching, which Einstein clearly studied carefully: there are handwritten highlights alongside almost a quarter of all the passages.[125] At the Albert Einstein Archives in Israel, I saw with my own eyes that this is more markings than any other book in Einstein's enormous personal library. Only a handful have highlights of any kind; most have no markings at all.[126]

Even more intriguing is Einstein's edition of the Tao Te Ching edited by Richard Wilhelm (1873–1930), one of the finest translators of the seminal works of Chinese spirituality.[127] Having spent twenty-five years in China, Wilhelm was not only fluent in the language, but steeped in the culture. Since the Tao Te Ching is a terse and poetic text, often impenetrable to Western readers, Wilhelm provided an ample commentary alongside his translation. And in his view, Lao Tzu had something in mind that bore more than a passing resemblance to Einstein's third phase. To begin with, Wilhelm credited Taoism with "the radical elimination of religious anthropomorphism"—that is, of purging the divine of human qualities.[128] But much like Einstein two millennia later, Lao Tzu had more in mind than mere disbelief. "He does not simply fight against popular religion," Wilhelm wrote, "but replaces it with something that is of a higher order, and leads further."[129] Remarkably, Wilhelm even pointed to the similarities he saw between Taoism and Spinoza's pantheism. Both Lao Tzu and Spinoza, he believed, considered all the infinitely diverse aspects of the universe as actually "identical in essence and different only in appearance."[130]

In struggling to express what the Tao really was, Wilhelm even suggested that "an analogy can be found in the laws of nature." As he explained it, "The laws of physics are expressed in all phenomena, but are not something distinct and separable," and "in the same way Lao Zi's Dao is present in all that happens."[131] Transcendent, omnipotent, and ultimately incomprehensible to mere human minds, the divine Tao was nonetheless utterly natural and immanent everywhere—"all-pervading," in Lao Tzu's words, and "omnipresent."[132] It's an idea that's almost perfectly analogous to Spinoza's Substance and Einstein's omnipresent "arch-force."[133]

With all this talk of the sacredness of creation, it might seem like the old religious dogmas are simply being rebranded under a new name. And in fact, pantheism is often confused with more traditional creeds that accept some kind of Creator. The easiest mistake to make is to conflate pantheism with Deism. Deism rose to prominence during the Age of Enlightenment as a kind of comforting compromise that made Christian faith compatible with the more critical modern mentality. Easily mocked ideas like miracles, divine revelation, and the literal truth of the Bible were dismissed in deference to the discoveries of science. But the basic belief persisted that a Creator God fashioned our universe with a purpose and a plan. From the Deist perspective, the orderly laws of physical existence and the miraculous organization of living beings provided incontrovertible evidence for God's existence and His goodness.[134] You don't hear the word *Deism* much these days, but the idea lives on among its intellectual descendants: creationism and intelligent design.

Although Einstein was often accused of atheism, it doesn't seem like anyone thought of him as a Deist during his own lifetime. But over the last couple of decades, this has become the dominant narrative defining his spirituality. One biographer has suggested that Einstein "settled into a deism" in later life and embraced a "middle-age deistic faith."[135] *Time* magazine, celebrating Einstein as its "Person of the Century," hailed him as "a philosopher with faith both in science and in the beauty of God's

Einstein's books on ancient Chinese philosophy, including a book on Confucius and several copies of the Tao Te Ching. He also owned the writings of the Taoist sage Chuang Tzu (not pictured). It's clear that he studied the German edition of the Tao Te Ching by Fiedler (bottom left) carefully: it contains more handwritten markings than any other book in his entire personal library, which totaled more than two thousand volumes. Einstein had great admiration for "the old Chinese sages" and endorsed their goal of creating "a community of free and happy human beings who by constant inward endeavor strive to liberate themselves." Richard Wilhelm, who edited and translated some of Einstein's volumes, suggested that the Tao could be understood in a way very similar to Einstein's arch-force. Just as "the laws of physics are expressed in all phenomena, but are not something distinct and separable," Wilhelm wrote, "in the same way Lao Tzu's Dao is present in all that happens." Photo © the author (2023), used with permission of the Albert Einstein Archives.

handiwork."[136] And Einstein has even been (mis)quoted as saying, "I believe in God; I have a very deep faith.... There's a spirit manifest in the laws of the universe...and to me that explains my faith in a Creator and a faith in God."[137]

But the devil's in the details, and the details here are dead wrong. Einstein never once declared any *faith* in a Creator or a personal God. On the rare occasions he used the F-word at all, he meant merely "the faith in the possibility that the regulations valid for the world of existence are rational, that is, comprehensible to reason."[138] In other words, he professed the fundamental *Pythagorean* faith, rather than any confidence in a Creator.[139]

Deism's wild leap of faith—from orderly cosmos to authoritarian Creator—was one Einstein was never willing to take. Although he often spoke of his "feeling of awe and reverence for the manifest Reason which appears in reality," in his vision of existence there was "no Will, nor Aim, nor an Ought, but only Being."[140] The existence of an "ordered regularity based on cosmic law" was obvious, but for Einstein this divine order didn't imply a Designer.[141] "To assume the existence of an unperceivable being," he insisted, "does not facilitate understanding the orderliness we find in the perceivable world."[142]

It's a subtle but essential distinction. For Einstein, the divine mind was not some faraway, phantomlike Creator. "The God Spinoza revered is my God, too," he insisted. "I meet Him every day in the harmonious laws which govern the universe."[143] Like Bruno and Spinoza, Einstein believed that "behind all the discernible concatenations" of tangible reality there was "something subtle, intangible."[144] But *behind* meant neither above nor beyond. Einstein understood this "infinitely superior spirit" to be *intrinsic* to existence.[145] "My God," he maintained, "appears as the physical world."[146]

And although much has been made of Einstein's occasional use of the word *God*, it's much ado about (almost) nothing. As one historian has pointed out, "All his remarks about thinking himself into God's mind occur in private letters or were said in personal conversations."[147] And in personal correspondence or private conversation, there was very little chance of misunderstanding: everyone who knew Einstein well understood that he had no sympathy for mainstream monotheism and used the word *God* merely metaphorically.[148] One of Einstein's old letters, recently made public, makes this plain. "The word God is for me nothing more than the expression and product of human weaknesses," he wrote. "No interpretation, no matter how elegant, can change this (for me)."[149] There was only a single sense in which Einstein ever accepted or understood such a word. "In common terms," he clarified, "one can describe it as 'pantheistic.'"[150]

And since the divine was immanent everywhere, there was no need to search the stars or ascend to Heaven in order to honor the Infinite. Einstein marked a passage in the Tao Te Ching that made precisely this point. "Without leaving home one may have the knowledge of the Universe," Lao Tzu claimed. "Without opening a window one may perceive the Divine Tao."[151] Einstein agreed. Ever since he'd seen a compass needle trembling in response to some tremendous hidden power, he'd known that "the force which controls our universe, within, and by which we have our being" filled all things, manifesting everywhere in the material world.[152] "Veneration for this force," he affirmed, "is my religion."[153]

And yet with so much wretchedness surrounding us, it's easy to argue that pantheism is precisely the kind of pathetic wishful thinking Einstein was so often accused of. To hold that a world filled with so much suffering is *holy* is a bold claim, and many people through the ages have dismissed it as blatantly paradoxical. Even Schopenhauer succumbed to this simplistic criticism, proclaiming that "pantheism is essentially and necessarily optimism"—and Schopenhauer was a pessimist with no patience for such a position.[154]

But it's a false paradox. Without even realizing it, Western culture has become deeply imbued with the idea that the divine can only be that which is detached from the world, standing outside of it as Creator and Judge. Determined to keep God at a safe distance from our allegedly imperfect world—and insistent that He could neither commit nor create evil—the wishful thinking of the Western mind has refused to grapple with the duality of the divine and hence has been unable to

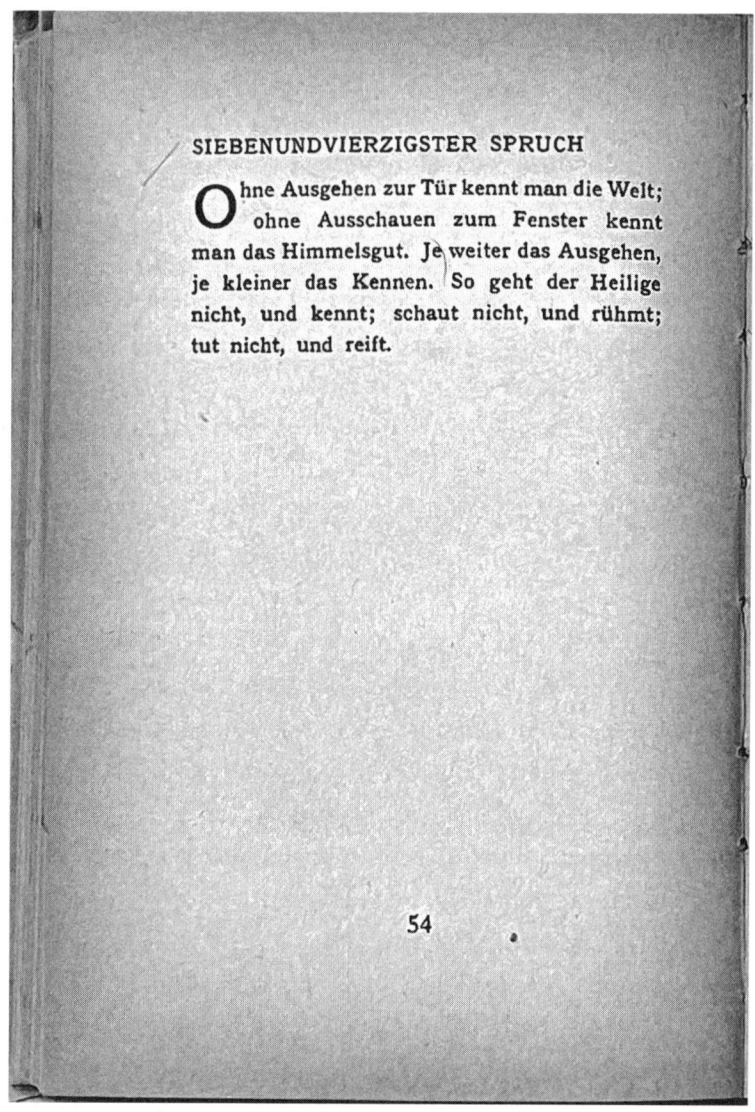

SIEBENUNDVIERZIGSTER SPRUCH

Ohne Ausgehen zur Tür kennt man die Welt; ohne Ausschauen zum Fenster kennt man das Himmelsgut. Je weiter das Ausgehen, je kleiner das Kennen. So geht der Heilige nicht, und kennt; schaut nicht, und rühmt; tut nicht, und reift.

54

Pictured above is one of the many passages Einstein marked by hand in his German edition of the Tao Te Ching. The distinctive diagonal pencil stroke was Einstein's characteristic way of marking a passage of special interest (compare with his marking of the passage from Democritus, in Chapter 4). Einstein made more handwritten markings in the Tao Te Ching than in any other book in his enormous personal library. In an English translation of the Tao Te Ching that Einstein owned, the above passage reads: "Without leaving home one may have the knowledge of the Universe. Without opening a window one may perceive the Divine Tao." Einstein probably marked this particular passage because it illustrates Taoism's close affinity with the pantheistic philosophers he admired, like Spinoza and Bruno. Photo © the author (2023), used with permission of the Albert Einstein Archives.

transcend it. The nuance of nondualism is lost on us, and the notion seems nonsensical.

But both Lao Tzu and Spinoza would warn us that it's a mistake to think of the cosmos-creating matrix the same way we think of God: as *good*. To be sure, they saw the divine substrate as abundant, creative, the mother of all things. But in the systems of Spinoza, Advaita, and Taoism, the womb of Being gives birth to both growth *and* decay. Darkness and death are as inseparable from its essence as light and life. Nowhere is this complementarity made more explicit than in the emblem of Taoism, the *taijitu* or yin-yang symbol, where light and dark are interdependent, containing and sustaining one another. For all these schools of thought, the divine force was generous but not benevolent, infinitely prolific but always impersonal. As Einstein put it, the "cosmic laws" couldn't be cajoled, bargained with, or "bribed by prayers or incense."[155]

And yet there was a reason Spinoza called his pantheist magnum opus the *Ethics*. Even though there was no God sitting in judgment of our actions, a pantheist perspective on reality still came with considerable ethical responsibilities. Although there was no way (and no need) to *obey* the Infinite, we could aim to *align* with the divine and attune ourselves to its august rhythms. Since all things, however small, had a place in the harmonious whole, tugging on even the tiniest thread of the great tapestry transformed the totality—and human beings were uniquely positioned to appreciate this interconnectedness. "It is only man's consciousness of what he does with his mind that elevates him above the animals," Einstein said, "and enables him to become aware of himself and his relationship to the universe."[156] And as the only creatures capable of understanding this interdependence, we had a hallowed responsibility to reflect this holy reality and embody its inner harmony in our outward actions. A cosmic religion required a cosmic human being, ready to live up to the demands of its divine birthright.

6

A HIGHER CALLING

Whatever there is of God and goodness in the universe,
it must work itself out and express itself through us.
What we do is of supreme importance to all humanity,
to history, to human destiny.

—Albert Einstein in conversation
with Algernon Black (1940)

A COSMIC RELIGION CALLED FOR A COSMIC CONSCIOUSNESS. AN AWAK-
ened mind could see that our immense cosmos was woven from
countless little threads—and could understand that this deep interdependence had important ethical implications. A unified cosmos entailed
a universal ethical system all people could follow, an ethos inspired by
interconnectedness. "The solution is simple enough," Einstein explained,
"and it seems also to echo from the teachings of the wise men of the past
always in the same strain: All men should let their conduct be guided
by the same principles."[1] And yet almost every ethical system through
history has instead divided humanity: faithful and infidels, saved and
damned, high-born Brahmins and unworthy untouchables.

Einstein's third phase demanded we abandon all these false divisions. It held that "humanity" was "one and undivided," and compelled us to consider every ethical conundrum from a cosmic perspective.[2] And even though Einstein was horrified by humanity's penchant for hatred and herd mentality, he held an almost heroic perspective on human potential. His hope was that every human being could one day live up to the demands implied by the immanent divine. "We must create a cosmic man, a man ruled by his conscience," he proclaimed, a person permeated by "a pure mind and pure intentions."[3] And in a world where nuclear weapons were widespread—the world in which we still live—the stakes could not possibly be higher. "We will be destroyed unless we create a cosmic conscience," Einstein feared. "And we have to begin to do that on an individual level."[4]

For Einstein, this call for unity and compassion was not just another arbitrary ethical injunction; it was the inescapable consequence of accepting that all was One. "To be just, noble, and benevolent is nothing but to translate my metaphysics into actions," as Schopenhauer once said. "All genuine virtue proceeds from the immediate and *intuitive* knowledge of the metaphysical identity of all beings."[5] Einstein's ethics likewise emerged from his own lived experience of the intimate links between all things. "As man becomes conscious of stupendous laws that govern the universe in perfect harmony he begins to realize how small he is. He sees the pettiness of human existence, with its ambitions and intrigues, its 'I am better than thou' creed. This is the beginning of cosmic religion within him," Einstein felt. "Fellowship and human service become his moral code."[6]

Einstein's lifelong love of peace was the most obvious manifestation of this metaphysical insight. Detractors often denounced Einstein's hopes for world peace as a perfect example of his hopeless romanticism. But Einstein's close friend Otto Nathan noticed long ago that it was a nuanced spirituality, rather than political naïveté, that inspired Einstein's ethics. "I know from the experience gained in the many years during which I enjoyed his friendship and confidence," Nathan wrote, "that, except for his devotion to science, no cause was more important or closer to his heart."[7] But Nathan wanted to know *why*. "How did this deep antipathy [to violence] originate?" he wondered. "What were the forces that inspired him?"[8]

Nathan wrestled with this question for years, and along the way he compiled a comprehensive collection of Einstein's pacifist writings.[9] In introducing this anthology, he admitted that there was, of course, a pragmatic side to Einstein's pacifism. According to Nathan, Einstein was "convinced that, as long as war existed as an accepted institution, the intellectual freedom of the individual—which he considered to be the foundation of human society—could not be realized. The existence of military institutions, the training of youth to serve as unthinking tools for the most flagrantly asocial purposes, the demoralizing effect of war preparations upon civilian life—Einstein considered all this incompatible with the dignity of free men."[10]

But Nathan also realized that "this rational analysis does not adequately explain Einstein's anxiety for a world in which man would live in total peace."[11] Nathan knew Einstein well, and knew he was "a deeply religious man—a religious unbeliever, as he once called himself—a man who revered nature with profound humility."[12] And he understood what hardly anyone else has ever grasped: that this secular spirituality was the secret source of Einstein's pacifist ideals. "I believe he must have felt that man's actions in war violated the sublime laws of the universe," Nathan wrote, "that the willful killing of millions interfered with nature's course for which he, the scientist, had the deepest reverence. It appears that *this* was the real source of his antipathy to brutality, the motive power behind his passionate, devoted efforts to help abolish the very institution of war."[13] With penetrating insight, Nathan noticed that "precisely because, as a scientist, he was engaged in the most abstract work of attempting to increase man's understanding of nature, he felt the compelling need to convince man not to flout nature's will."[14]

Einstein himself confirmed Nathan's conjecture. Everywhere in his writings, he proclaimed that pacifism wasn't just a matter of practical necessity, but part of a sweeping spiritual perspective that permeated his whole being. "My pacifism is an instinctive feeling, a feeling that possesses me," he once wrote. "My attitude is not derived from any intellectual theory but is based on my deepest antipathy to every kind of cruelty and hatred."[15] Einstein therefore saw it as our "solemn and transcendent duty to do all in our power" to prevent war and promote peace.[16] He even went as far as to say that "whoever recognizes this clearly, but fails to orient his political attitude accordingly, has no right to consider himself a religious person."[17] In other words, not only was pacifism the essence of Einstein's own spirituality; in his view, it was the prerequisite of any real religion.

There's every indication that this was an inborn inclination that had been with Einstein all his life. Even as a young boy, he'd once started crying while watching a military parade march past the family home in Munich. Other children ran outdoors to join the marching soldiers, but Einstein, with tear-filled eyes, told his parents that "when I grow up, I don't want to be one of those poor people."[18] This antipathy only increased as he grew older. As an adolescent, he was appalled at how German schools were used "for inculcating ideals of imperialistic power and military success."[19] By the time he was sixteen, his revulsion for Germany's rampant militarism led him to take the radical step of renouncing his citizenship entirely.[20] "This heroism at command, this senseless violence, this accursed bombast of patriotism—how intensely I despise them!" he wrote.[21] And when the First World War finally broke out, he was one of only a tiny handful of German intellectuals who were willing to openly oppose it.[22]

But Einstein wasn't alone. Even during the tumultuous twentieth century, which bore witness to the two most horrendous wars in human history, there were a rare few who felt the full fervor of pacifism and refused to participate in war on principle: the so-called conscientious objectors. Most of these war resisters were publicly maligned: their motives were misunderstood as cowardice and many were imprisoned for adhering to their ideals. But Einstein, appalled by the public's animosity, defended them against the world's prejudice in both private letters and public pronouncements. "Men who, by their religious and moral convictions, are constrained to refuse military service should not be treated as criminals," he wrote. "Nor should anyone be permitted to sit in judgment on the question of whether such a refusal is rooted in deep conviction or in less worthy motives."[23]

It was not cowardice, Einstein concluded, but an authentic religious feeling that forced such people to practice pacifism—even in spite of its widespread unpopularity. "The conscientious objector is a revolutionary," he wrote.[24] These were "people who would rather submit to punishment and social ostracism than act against their conscience. The existence of such a moral elite is a prerequisite to any fundamental change."[25] Einstein even compared contemporary conscientious objectors to the religious martyrs of the past. "Is this persecution," he asked, "any less shameful for society than the persecution to which the religious martyrs were subjected in earlier centuries?"[26]

Einstein eventually came to see conscientious objection as a program that could be applied on a global scale. Perhaps the climax of his public career as a pacifist came in December 1930, when he gave an infamous speech calling for people of all nations to categorically reject compulsory military service. Hitler's National Socialist party had made major strides in the federal election earlier that year and had started enacting anti-Semitic policies in Germany; the first shadows of the Second World War were already starting to darken the skies over Europe. But despite these dire developments, Einstein declared that "the true pacifist must refuse military duty," and he urged "uncompromising war resistance and refusal to do military service under any circumstances." He anticipated that some might see this as an empty gesture, but he disagreed. "The timid may say, 'What is the use? We shall be sent to prison.' To them I would reply: Even if only two per cent of those assigned to perform military service should announce their refusal to fight . . . governments would be powerless, they would not dare send such a large number of people to jail." The address became known as Einstein's "Two-Percent Speech" and established his enduring reputation as a pacifist in Gandhi's vein: "not to use violence in fighting for our cause," as Einstein explained, but by "nonparticipation in what we believe is evil."[27]

But even though pacifism was a constant throughout his long life, Einstein's views on peace were variable. When it came to the overwhelming inhumanity of Adolf Hitler, Einstein equivocated. In an interview given in 1935, although he admitted that "I admire Gandhi greatly," he worried that there were "weaknesses in his program." Contra Gandhi, Einstein now contended that absolute nonviolence could be "practiced only under ideal conditions... It could not be carried out against the Nazi Party today."[28] It was a major reversal of what he'd recommended in his "Two-Percent Speech" just five years earlier, but his views were largely vindicated by the events that followed. During a conversation with the poet and pacifist William Hermanns in 1943, with the war raging, Einstein summed up the new situation. "Things have changed," he said. "A dictator has come who thrives on the passive attitude of the masses.... In the face of this new threat to humanity, it would be suicide to advocate nonresistance.... He who has witnessed the atrocities of the Hitler Reich has no choice but to arm. Right now I would be the first to take up arms if I saw my dear ones being enslaved or killed and our civilization being wiped out."[29]

Hermanns remembered being "gripped by a feeling of awe for this man's courage in admitting his about-face so frankly and publicly," and he told Einstein that "your change is a shock for many." As Hermanns recalled: "Einstein turned to me with a mystic's smile. 'Yes, I know, I'm no longer an unconditional pacifist, but a realistic one... I'm fully aware that I made many enemies when I changed my mind about pacifism. I displeased the Quakers, of course, as well as the followers of Bertrand Russell, and of Gandhi. But principles are made for men and not men for principles.'"[30]

Years after the war, Einstein elaborated on his position in an exchange of letters with the Japanese pacifist Seiei Shinohara. "While I am a convinced pacifist," he explained, "there are circumstances in which I believe the use of force is appropriate—namely, in the face of an enemy unconditionally bent on destroying me and my people.... This is why I believe that the use of force was indicated and justified in the case of Nazi Germany."[31] Shinohara wasn't impressed. He rebuked Einstein for signing the letter to Roosevelt encouraging research on an atomic bomb and insinuated that Einstein shared responsibility for the immolation of countless innocent civilians in Hiroshima and Nagasaki. "I have always condemned the use of the atomic bomb against Japan," Einstein replied, a little rancorously. "However, I was completely powerless to prevent the fateful decision for which I am as little responsible as you are for the deeds of the Japanese in Korea and China."[32] As disappointing as Einstein's regression to a more realistic pacifism was to the purists, it's to his credit that he candidly admitted his reversal and clearly explained his reasons. Bitter experience had taught him that the world was not yet ready for absolute pacificism. A German Jew who lived in the shadow of National Socialism can perhaps be forgiven for falling a little short of such lofty ideals.

And if Einstein ever harbored any lingering regrets, Gandhi's assassination in 1948 provided a poignant reminder of the risks posed by a strict commitment to pacifist principles. "Everyone concerned with a better future for mankind must be deeply moved by the tragic death of Gandhi," Einstein said at a memorial service. "He died a victim of his own principle, the principle of nonviolence. He died because, in a time of disorder and general unrest in his country, he refused any personal armed protection." Einstein deeply admired the Mahatma's "unshakable belief that the use of force is an evil in itself," but he himself had no desire to end up a martyr to the cause.[33]

Although Einstein abandoned absolute pacifism in human affairs, his ethics actually became increasingly *inclusive* as his ideas evolved over the years. In the latter half of his life, his notion of nonviolence came to include not just human beings, but *all* forms of life. He let it be known that "the love of living creatures is for me the finest and best trait of mankind," and his home in Princeton became a minor menagerie, hosting a tomcat named Tiger, a dog named Chico Marx, and even a parrot by the name of Bibo.[34] His affection for the animals is obvious from surviving anecdotes. "The dog is very smart," he once said. "He feels sorry for me because I receive so much mail; that's why he tries to bite the mailman."[35] And when Tiger seemed depressed because pouring rain was keeping him cooped up in the house, Einstein was overheard telling him, "I know what's wrong, dear fellow, but I don't know how to turn it off."[36]

By no means was Einstein's love for animals limited to a fondness for his household pets. On the contrary, he considered "kindness toward all creatures" to be a necessary next step on the path of moral progress, and insisted that we should "educate the youth in the spirit of... love toward all living creatures."[37] Eventually Einstein decided to embody these values directly and adopt a vegetarian diet. We don't know exactly when he stopped eating meat, but we do know from his letters that before he died he'd made the switch. "So I am living without fats, without meat, without fish, but am feeling quite well this way," he wrote to a friend in 1954. "It almost seems to me," he mused, "that man was not born to be a carnivore."[38]

As usual, Einstein was ahead of his time. Vegetarianism is on the rise today, but it was an incredibly rare practice in postwar America. And yet even in this uncomprehending era, Einstein obeyed his conscience and put his strange principles into practice. Inexplicably, some Einstein scholars still deny that this was a meaningful gesture. Even one of the world's leading experts has insisted that "Einstein was probably not a vegetarian by choice, for he left behind no remarks that it was a moral issue for him."[39] But Einstein openly admitted that "I have always eaten animal flesh with a somewhat guilty conscience," and he couldn't have made it more clear that this *was* a moral matter for him.[40] "Although I have been prevented by outward circumstances from observing a strictly vegetarian diet, I have long been an adherent to the cause in principle," he acknowledged as early as 1930. "Besides agreeing with the aims of vegetarianism for aesthetic and moral reasons, it is my view that a vegetarian manner of living... would most beneficially influence the lot of mankind."[41] And there's no doubt that deep spiritual feelings drove this decision. "Life is sacred," as Einstein once explained. "It is the supreme value, to which all other values are subordinate."[42]

For Einstein, this wasn't just another arbitrary religious rule meant to restrict our actions. On the contrary: rather than reining us in, a reverence for life freed us from the rigid restraints of our own selfishness. Einstein saw the smallness of our ego-centered selves as "a kind of prison for us, restricting us to our personal desires and to affection for a few persons nearest to us." Real ethics began only when and where the ego ended. "Our task must be to free ourselves from this prison," Einstein said, "by widening our circle of compassion to embrace all living creatures."[43] And in Einstein's eyes, service to the whole sacred spectrum of life was ultimately what made our own limited existence meaningful. "The life of the individual," he insisted, "has meaning only insofar as it aids in making the life of every living thing nobler and more beautiful."[44]

Since this all-inclusive notion of nonviolence was the essence of Einstein's ethics, it's essential that we understand how he came to see all life as sacred. There are some who think that Judaism was the source. One scholar has argued that there was a "biblical precedent for Einstein's idea of the expansion of selfhood as a response to suffering," and that "Einstein's own path to a wider self was through his Jewishness."[45] Another scholar, a biographer who knew Einstein well, has alleged that "it was in the realm of ethical relations that Einstein felt most deeply his Jewish heritage," because "the moral code of the Hebrews...is based upon the principle that life is holy."[46]

But no one who's actually read the Jewish Bible can honestly claim that it advocates nonviolence. Animal life certainly isn't seen as sacred: the ritual murder of living creatures for religious purposes is pervasive throughout the Old Testament. God often asks for animal sacrifices, and the Hebrews almost always comply.[47] Human beings don't fare much better: the death penalty is doled out for everything from adultery to worshipping the wrong god to working on weekends.[48] And even mass murder is celebrated on several occasions: the God of the Old Testament is just fine with genocide, as long as it happens to enemies of the Jewish people.[49] Moses does give the commandment "Thou shalt not kill," but all he meant was *don't kill other Jews*—unless they break one of God's many laws and deserve the death penalty. And of course, the punishment for committing the ultimate crime, murder, was (you guessed it) death.[50]

With Jesus and the New Testament, there's no denying that the Bible has taken a decisive turn toward gentler moral injunctions.[51] Jesus even preached a kind of absolute pacifism in telling us to turn the other cheek.[52] But neither Christianity nor any other monotheistic religion ever extended the circle of compassion beyond human beings. In fact, the only Western religious movement that ever promoted compassion toward animals as a major moral principle was the school of that peculiar man who keeps popping up in the story of Einstein's spirituality: Pythagoras. In fact, the name of Pythagoras was so closely associated with abstention from eating animals that for centuries vegetarianism was known in the Western world as "the Pythagorean diet."[53]

But as soon as we look beyond the Western tradition, the origin of Einstein's ideals becomes obvious. Because in the East, a reverence for life has formed the foundation of almost all religions for thousands of years. Indeed, the all-encompassing ethos Einstein eventually embraced and tried so hard to communicate is nothing other than ahimsa, the ancient Indian ideal of nonviolence. And although Einstein never used the word *ahimsa* himself, he endeavored to live his life according to the Latin equivalent, insisting that "we must above all abide by the old medical maxim: *Non nocere* [Do no harm]."[54]

The origin of the ideal of ahimsa is lost in legend, but it's obviously an ancient idea, one that no existing tradition can claim as its own exclusive innovation. Some scholars argue that ahimsa appears in India's oldest sacred scriptures, the Vedas.[55] Circumstantial evidence has led others to speculate that the idea goes back even further, to the ancient Indus Valley civilization.[56] But probably the first unequivocal use of the term as we understand it today is in the Chandogya Upanishad.[57] And even in this ancient text, nonviolence is already linked with nirvana. The sage "who practices non-hatred [ahimsa] to all creatures," the Chandogya Upanishad claims, "reaches the Brahma-world [and] does not return hither again."[58] From the very beginning, then, ahimsa was presented as an essential part of the path to enlightenment.

Whatever its origins, ahimsa was enthusiastically embraced by all the subsequent Indian religious traditions familiar to us today. It's the first of the five precepts followed by Buddhists, and it's so fundamental to various forms of Hinduism that almost a third of India's population, some 300 million people, are vegetarians today. But ahimsa really reached its apex in a tradition that's almost unknown outside India. Jainism is one of the world's smallest religions, boasting only a few million followers. It was founded around twenty-five hundred years ago by a man known as Mahavira. An older contemporary of the Buddha, Mahavira's life story shares many similarities with the much better-known tale of Siddhartha.* Both were born in the same northeastern region of India, both grew up as princes in royal households, and both left home to live the religious life around the age of thirty. Both are said to have attained enlightenment after years of arduous spiritual practice and then roamed around India for the remainder of their lives teaching disciples.

What distinguishes Jainism is that, even more than Hinduism and Buddhism, Mahavira made ahimsa an absolute—and his emphasis on nonviolence above all other values continues to have an immense influence on his followers. Today, over 90 percent of ordinary Jains are vegetarians, and the more hardcore adherents take ahimsa to astonishing extremes.[59] Jain monks and nuns often wear a cloth face mask to avoid accidentally inhaling any insects, or carry a sacred communal broom, called a *picchi*, to brush bugs out of harm's way when walking someplace or sitting down to meditate. Although the Jains might not have been the first to advocate ahimsa, undoubtedly they were (and still are) the greatest virtuosos of nonviolence.

Einstein might have encountered ahimsa in any number of places. But there's little doubt that he was exposed to religious arguments for animal rights when he read Schopenhauer's *Parerga and Paralipomena* in his early twenties.[60] Intriguingly, the idea appears in Schopenhauer's

---

* This was the Buddha's given name.

discussion of the ancient Greek philosopher Empedocles (yet another Pythagorean).[61] According to Schopenhauer, the essence of Empedocles's teaching was that our "souls were once in a state of infinite bliss and have fallen into present ruin through their own fault and sin." But there was a way back to this heavenly bliss. "Through virtue and moral purity," including "the abstention from an animal-based diet," our souls could "return to their former state." Schopenhauer made sure to underline the obvious affinity with Eastern religions, arguing that "this ancient Greek was aware of the same original wisdom that makes up the fundamental idea of Brahmanism [i.e., Hinduism] and Buddhism."[62]

Schopenhauer admired these ideas enormously. He made ahimsa the mainstay of his own morality and took Western philosophers to task for thoroughly ignoring animals when formulating their ethics. "The unpardonable forgetfulness in which the lower animals have hitherto been left by the moralists of Europe," he complained, "is well known. It is pretended that the beasts have no rights. They persuade themselves that our conduct in regard to them has nothing to do with morals or (to speak the language of their morality), that we have no duties towards animals; a doctrine revolting, gross, and barbarous." And it's obvious that Schopenhauer saw Eastern religions as offering a more appealing alternative. He even insisted that "I know of no more beautiful prayer than that which the Hindus of old used in closing their public spectacles... 'May all that have life be delivered from suffering!'"[63]

Einstein put this ancient prayer into practice in his personal life, but the "Hindus of old" were long gone, living on only in legends, and it wasn't enough for him to honor the teachers and teachings of the past. Einstein was looking for living avatars of ahimsa, people whom he could hold up to the human race as contemporary exemplars of the cosmic human being. There were two people in particular whom he considered contenders for this crown, and both of these paragons of peace traced the origin of their all-embracing nonviolence to Jainism.

The first was Mahatma Gandhi. Born Mohandas Gandhi in 1869, he studied law in London before moving his practice to South Africa and gaining notoriety for demanding equality for Indian citizens in the British colonies. He went on to world fame when he became the leader of the nonviolent Indian Independence Movement, which succeeded in securing India's sovereignty in 1947. Einstein heaped praise on this apostle of peace, calling the Mahatma "a man of wisdom and humility, armed with resolve and inflexible consistency, who has devoted all his strength to the uplifting of his people and the betterment of their lot; a man who has confronted the brutality of Europe with the dignity of the simple human being, and thus at all times risen superior."[64]

Always on the lookout for actual exemplars, Einstein especially admired how Gandhi was an embodiment of his own ethical principles, a "living

Einstein in his study at his home in Princeton in 1953. Note the portrait of Gandhi on the wall behind him. The only other non-scientist accorded such an honor was the philosopher Arthur Schopenhauer. Mohandas Gandhi (1869–1948) was renowned around the world as an apostle of ahimsa (nonviolence). Although the two men never met in person, Einstein held Gandhi in higher esteem than any other living human being. "Generations to come," he gushed, "will scarce believe that such a one as this ever in flesh and blood walked upon this Earth." Einstein owned many books by or about Gandhi and claimed that careful study of these writings had given him "the greatest admiration for Gandhi and for the Indian tradition in general." Photo © Alan Richards (1953), used with permission of the Institute for Advanced Study.

example of a morally exalted way of life."[65] On Gandhi's seventy-fifth birthday, Einstein gushed that "generations to come, it may well be, will scarce believe that such a one as this ever in flesh and blood walked upon this Earth."[66] A portrait of Gandhi even hung in Einstein's home in Princeton—and we know that only a handful of his greatest heroes were ever accorded such an honor.[67]

Much like Einstein, many people revered Gandhi as an almost god-like figure, and this veneration is manifest in the moniker by which he's remembered today: Mahatma, which means "great soul."[68] But Mohandas Gandhi wasn't born a mahatma; he became a great ethical exemplar one little step at a time. And Einstein must have known all about how he'd made the extraordinary transition from mere mortal to moral titan. "I have studied the works of Gandhi," Einstein wrote, "with real admiration."[69] His personal library contains almost a dozen books on or by Gandhi, and Einstein claimed that exploring this oeuvre had given him "the greatest admiration for Gandhi and for the Indian tradition in general."[70] Clearly, Einstein was conscious that Gandhi's spiritual roots were planted in the soil of India. He even objected when someone suggested that Christian pacifists might have been Gandhi's primary inspiration. "It may well be that Thoreau has in some way influenced Gandhi's thought," Einstein replied. "But it should not be forgotten that Gandhi's development resulted from extraordinary intellectual and moral forces in combination with political ingenuity and a unique situation. I think Gandhi would have been Gandhi even without Thoreau and Tolstoi."[71]

And in fact it's obvious from Gandhi's own writings that his inspiration came mainly from spiritual mentors who were much more ancient—and much closer to home. Gandhi grew up in the Indian state of Gujarat, which contains one of the country's highest concentrations of Jains. In his autobiography (which Einstein of course owned), Gandhi recalled that "Jainism was strong in Gujarat, and its influence was felt everywhere and on all occasions. The opposition to and abhorrence of meat-eating that existed in Gujarat among the Jains... [was] to be seen nowhere else in India or outside in such strength. These were the traditions in which I was born and bred."[72]

Gandhi himself was not a Jain, but he had a deep respect for their religion.[73] "I know the Jains," he said. "I know them and the principles of their religion intimately. . . . I have learned much from their books. My contact with many Jain friends has helped me much."[74] And Jainism's "most remarkable characteristic," he thought, "was its scrupulous regard for all things that lived."[75] Gandhi put these ancient Jain principles into stunningly effective practice in the modern world, and Einstein had the utmost respect for the results. "Gandhi," Einstein said, "gave proof of what sacrifice man is capable once he has discovered the right path." For Einstein, Gandhi's exceptional life and immense political achievements bore "living testimony to the fact that man's will, sustained by an indomitable conviction, is more powerful than material forces that seem insurmountable."[76] Einstein even felt that we were "fortunate and should be grateful that fate has bestowed upon us so luminous a contemporary—a beacon to the generations to come."[77]

But there was one other person Einstein felt was on par with his highest hero—"the only Westerner," as he put it, "who has had a moral effect on this generation comparable to Gandhi's."[78] This was the physician, theologian, and pacifist Albert Schweitzer. Schweitzer's whole philosophy focused on finding a universal ethical framework for the modern world, and he referred to his fundamental conviction as "Reverence for Life." The essence of his ethics was simple yet sublime. "Good consists in maintaining, promoting, and enhancing life," Schweitzer said, whereas "destroying, injuring, and limiting life are evil."[79] He practiced what he preached. He spent some three decades as a doctor treating the horrific consequences of tropical diseases in West Africa, where he was also one of the harshest critics of the crimes of colonialism. Einstein called him "a great figure who bids for the moral leadership of the world," and Schweitzer's untiring humanitarian work ultimately won him the Nobel Peace Prize in 1952.[80]

As a theologian, Schweitzer was a diligent student of Western scripture and wrote many books about Jesus and Christianity. But in the latter half of his life he also composed a book on the East, *Indian Thought and Its Development*. On the whole, Schweitzer didn't have much sympathy for Eastern religion, which he saw as overly passive and pessimistic. But in a chapter on Jainism, he gave credit where credit was clearly due. "The laying down of the commandment to not kill and to not damage is one of the greatest events in the spiritual history of mankind. Starting from its principle," he speculated, "ancient Indian thought...reaches the tremendous discovery that ethics knows no bounds. So far as we know, this is for the first time clearly expressed in Jainism."[81] And so Einstein, through his idols Mahatma Gandhi and Albert Schweitzer, had absorbed the highest ethical instructions that India had to offer—and he echoed and honored these ancient ideals in his own ethics.

Of course, Einstein well understood that it was one thing to *identify* ideal ethical behavior; it was another thing entirely to entice people to actually put these ideals into practice. "The real difficulty," Einstein realized, "the difficulty which has baffled the sages of all times, is rather this: how can we make our teaching so potent in the emotional life of man, that its influence should withstand the pressure of the elemental psychic forces in the individual?"[82] Century after century, the same problem confronted every spiritual system seeking to elevate humanity: how to convince people to *choose* a higher path, rather than coerce them into embracing it out of fear; how to persuade people to see the improvement of society not as an onerous duty, but rather, in Einstein's words, "to look upon social problems as so many opportunities for joyous service towards a better life."[83] The perennial problem, Einstein realized, was to help people see that "moral conduct does not mean merely a stern demand to renounce some of the desired joys of life, but rather a sociable interest in a happier lot for all men."[84]

It went without saying that neither rational arguments nor humiliating harangues would win people over to ethical ideals like ahimsa. Such methods had failed for thousands of years. Gandhi gave considerable thought to the conundrum of how to motivate real and lasting moral change, and he came to a surprising conclusion: personal communion with the powers of the cosmos was the surest path to a committed pacifism. "Non-violence does not work in the same way as violence," he explained. "It works in the opposite way. An armed man naturally relies on arms. A man who is intentionally unarmed relies upon the unseen force called God by poets, but called the unknown by scientists."[85] Only by becoming attuned to this absolute reality, Gandhi believed, could we come to embody the ultimate ethical ideal.

Einstein came to the same conclusion. "My God may not be your idea of God," he once told a friend, "but one thing I know of my God—he makes me a humanitarian."[86] For Einstein, nonviolence was the natural corollary of nondualism; compassion was the natural consequence of wisdom. The same sentiment is expressed in a passage he highlighted in his copy of the Tao Te Ching. "The Truly Wise dwell in the hearts of all beings," it reads; they "treat the world compassionately; their hearts are all-encompassing."[87]

For Einstein, this wasn't just some mystical mumbo-jumbo. On the contrary, a feeling of empathy and interconnectedness with other animate beings was precisely what remained after *purging* religion of all its associated nonsense. A "sympathetic feeling in joy and in sorrow," as Einstein explained, "is that which is left of religion when it has been purified of the elements of superstition."[88] To embrace ahimsa, then, was simply to put into practice what he knew from his own highest personal experiences: everything was One.

7

A CONSCIOUS FREEDOM

Only if outward and inner freedom are constantly and consciously pursued is there a possibility of spiritual development.

—Albert Einstein, "On Freedom" (1940)

E INSTEIN UNDERSTOOD THAT WE WERE BOTH ENNOBLED AND IMPRIS-oned by the Infinite. Becoming aware of our interconnection with everything everywhere broadened our sense of ourselves enormously, of course. But it also implied that we were inextricably intertwined with all other things across the endless stretches of space—and unwitting heirs of every event that had ever occurred over incomprehensible spans of time. All things, including even our individual minds, were woven together in a seamless, self-consistent tapestry.

To Einstein, the notion of true autonomy within such an integrated totality was nonsensical. For all our pretensions of individuality and agency, there could be no fleeing from physical law, no escaping our evolutionary heritage. And so despite all his calls for heroic ethical action, he denied that we had free will in any fundamental sense. "Everything is determined, the beginning as well as the end, by forces over which we have no control," he once said. "It is determined for the insect as well as for the star. Human beings, vegetables, or cosmic dust, we all dance to a mysterious tune."[1]

Einstein's position seems deeply paradoxical. He claimed we had no control over our actions, but he called on courageous individuals to inspire others to act more nobly. He insisted on the unbreakable necessity of natural law, but he pleaded with us to change the course of history and build a more peaceful world. He held that humans had no real moral responsibility, but he summoned us to create our own high ethical standards—and live up to them. It seems as if Einstein's ideas are irreconcilable, if not downright contradictory.

And yet deep down he knew we were not merely the playthings of external powers. Einstein understood that every interaction was reciprocal—even that between one little conscious being and the rest of the cosmos. It was true that we were enmeshed in the vast ecosystem of existence. But inside every one of us was a small seed of autonomy that, properly cared for, could grow and flower into an authentic "freedom of the spirit."[2] As Einstein saw it, the fact that our incipient autonomy required conscious cultivation was a blessing rather than a burden—an exceptional opportunity to show the wider network of Reality the true nature, and real nerve, of the little node of self.

Einstein's implacable insistence on determinism might seem like a stubborn superstition, but actually it was an inevitable consequence of his commitment to a unified reality. It wasn't some bleak vision of being imprisoned by unbreakable laws, but rather an uplifting intuition of being interwoven with all things. "Life is a great tapestry," he once said. "The individual is only an insignificant thread in an immense and miraculous pattern."[3] A human being might be an especially exquisite part of the tapestry of existence, but it was a part all the same. And if all things were really One, then the thesis of free will was fundamentally flawed.

An obvious implication was that we had no responsibility for our actions, good or evil, and Einstein defended exactly this position. "A man's actions are determined by necessity external and internal," he wrote. "He cannot be responsible, any more than an inanimate object is responsible for the motions it undergoes."[4] Elsewhere, he made his position even more plain: "Human beings in their thoughts, feelings and actions are not free agents, but are as subject to the inexorable laws of cause and effect as are the stars in their courses."[5]

Einstein's understanding of determinism ran deep, and the standard view of causality was too simplistic and too superficial to do it justice. "Much of the misunderstanding encountered in all this question of causation," he contended, "is due to the rather rudimentary formulation of the causal principle which has been in vogue until now."[6] The rudimentary formulation he had in mind was the reductionist "billiard ball" model of causality most of us are familiar with: one moving ball (the cause) hits another that was previously at rest and makes it move (the effect). It's an appealing picture, but ever since Isaac Newton, physicists have acknowledged that cause and effect are simply two sides of one *simultaneous* event. It certainly seems to us that one ball hits the other and makes it move, directly causing its motion. But from the physical standpoint, they're merely two interacting systems exchanging energy, neither of which is privileged as causal agent.

The critique of billiard ball causality is a little abstract; the problem isn't always apparent. But it's easy to see how the classical conception of causality falls apart under more complex conditions. Consider our solar system. To say that the Sun is the *cause* of the Earth's orbit is a convenient, and very accurate, approximation. Strictly speaking, though, it isn't true. *Both* bodies exert an effect *on each other*—and *at exactly the same time*. The Sun is so much more massive than the Earth (it's got about a million-to-one advantage) that we can comfortably ignore the pull of our little planet. But many other solar systems happen to have two stars of roughly equal mass at their center. Who orbits whom in these binary star systems? We can model the movements of both stars, but the question of causality barely makes sense. By the time we reach the infamous three-body problem (three masses mutually interacting), physicists are already at a complete loss. And if we try even for a moment to grasp the endless multitude of apparently independent objects affecting one another all over the cosmos, the mind quickly gives up—and so does science.

Spinoza made a similar point centuries earlier in defense of his own determinism. "Let us imagine, if you please, a tiny worm living in the blood," he wrote. "That worm would be living in the blood as we are living in our part of the universe . . . and it could have no idea as to how all the parts are controlled by the overall nature of the blood and compelled to mutual adaptation as the overall nature of the blood requires." In other words, the worm's local perspective inevitably limited its appreciation of the larger picture. "Now all the bodies in Nature can and should be conceived in the same way," Spinoza insisted. From hydrogen atom to human agent, "every body must be considered as a part of the whole universe, and as agreeing with the whole and cohering with the other parts."[7]

These aren't idle analogies; they're accurate descriptions of our total inability to tease apart the infinity of mutual interactions that makes up our universe. And it was this mind-boggling mutuality that drove Einstein to embrace determinism. He wasn't saying human beings lacked freedom because we were just bodies being bounced around like billiard balls in Newton's mechanical clockwork universe. He was pointing out, instead, that all individuality and all independence were ultimately illusory. Dividing things up into distinct objects with cause-effect relationships was convenient, and in fact necessary, for us to function in the world. But viewed philosophically, individual objects interacting through fields of force was a simplistic superstition.

Never content with mere theory, Einstein wanted to demonstrate the physical reality of this philosophical principle. "Can we think of matter and field as two distinct and different realities?" he once asked. "Given a small particle of matter we could picture in a naïve way that there is a definite surface of the particle where it ceases to exist and its gravitational field appears. . . . But what are the physical criterions distinguishing matter and field?"[8]

Einstein's answer was alarming: *there are no such criteria*. Einstein had already shown with his famous formula $E = mc^2$ that mass and energy were equivalent; all existence could therefore be conceived of as a single energetic substrate manifesting in so many different forms. What Einstein realized was that the incredibly uneven distribution of this single substrate led to the irresistible illusion that our world was made up of independent, interacting parts. "By far the greatest part of energy is concentrated in matter," Einstein pointed out, "but the field surrounding the particle also represents energy, though in an incomparably smaller quantity." All the same, Einstein insisted that "the difference between matter and field is a quantitative rather than a qualitative one." And he decided that "the division into matter and field" was "something artificial and not clearly defined."[9]

Both the ambitions and the implications of this perspective were enormous:

> Could we not reject the concept of matter and build a pure field physics? What impresses our senses as matter is really a great concentration of energy into a comparatively small space. We could regard matter as the regions in space where the field is extremely strong. In this way a new philosophical background could be created. Its final aim would be the explanation of all events in nature by...laws valid always and everywhere. A thrown stone is, from this point of view, a changing field, where the states of greatest field intensity travel through space with the velocity of the stone. There would be no place, in our new physics, for both field and matter, field being the only reality.[10]

Einstein's pure field physics was a complete reconceptualization of the alleged independence of all objects. Every separate particle, every person finding their way on Earth, every planet floating silently through fathomless space—in Einstein's opinion, all were woven from a single shimmering fabric.

And Einstein was just getting warmed up. Even with this newfound appreciation for the integration of all things in space, we'd still forgotten all about *time*. With general relativity, Einstein had established the synthesis of space and time as a single unified *spacetime*, but classical conceptions of causality completely ignored the deep past and distant future. Einstein decided it was time to change all that. "I believe that events in nature are controlled by a much stricter and more closely binding law than we suspect today when we speak of one event being the *cause* of another," he explained. "Our concept here is confined to one happening within one time-section. It is dissected from the whole process. Our present way of applying the causal principle is quite superficial."[11]

Einstein elaborated on his position using a musical metaphor that no doubt would have pleased the Pythagoreans. "We are like a juvenile learner at the piano, just relating one note to that which immediately precedes or follows. To an extent this may be very well when one is dealing with very simple and primitive compositions; but it will not do for the interpretation of a Bach fugue."[12] Einstein's intuition told him that we inhabited a harmonious whole in which every particular melody had its complementary counterpoint. Every little morsel of matter and every ephemeral moment of time was related to every other. Nothing existed, or endured, in isolation. For Einstein, the conclusion was clear: "We must further enlarge and refine our concept of causality."[13]

And it wasn't just our intimate interconnections with the spatiotemporal fabric of the physical world that fettered our freedom; it was also the inexplicable impulses influencing us from within. "We are driven by some inner force which we are not always aware of," Einstein said, "so I hesitate to talk about free will."[14] Over and over, he emphasized the importance of inborn inclinations of which we knew almost nothing—and over which we had almost no control. "We all know," Einstein said, "that our conscious acts spring from our desires and our fears. Intuition tells us that this is true also of our fellows and of the higher animals.... All these primary impulses, not easily described in words, are the springs of man's actions. All such action would cease if those powerful elemental forces were to cease stirring within us."[15] And there was no way to bypass our bondage to the instinctual behaviors inherent in all biological beings.

"For the most part," Einstein pointed out, "I do the thing which my own nature drives me to do."[16]

Today, we tend to think of an emphasis on unconscious urges and ancient instincts as a Freudian theory, but the essential idea arose centuries earlier.[17] One of its most forceful and thoughtful exponents was none other than Arthur Schopenhauer. Now known to have been a major influence on Freud, Schopenhauer likewise conceived of the conscious mind as mostly powerless, a thin film of awareness atop an ancient infrastructure of invincible instincts and intrinsic modes of thought.[18] Underlying it all was a mysterious, monolithic metaphysical entity Schopenhauer called the Will. From the humblest fragment of humdrum matter to the highest thoughts of the human mind, everything emanated from this unfathomable force. "Hence we get the strange fact that everyone considers himself to be *a priori* quite free, even in his individual actions, and imagines he can at any moment enter upon a different way of life," Schopenhauer mused. "But *a posteriori* through experience, he finds to his astonishment that he is not free, but liable to necessity; that notwithstanding all his resolutions and reflections he does not change his conduct."[19] For Schopenhauer, the innermost source of even our most intimate thoughts was alien and inscrutable, and Einstein agreed with his idea that an all-powerful Will ruled us from within. "I believe with Schopenhauer," he said, "we can do what we wish, but we can only wish what we must."[20]

As usual, Einstein was ahead of the curve in pointing to the biological basis for our otherwise inexplicable moods and mental states—and he didn't spare himself the humiliation of having little to no freedom. "My own career was undoubtedly determined, not by my own will, but by various factors, over which I have no control," he claimed in an interview, "primarily those mysterious glands in which nature prepares the very essence of life, our internal secretions."[21] The interviewer, intrigued, informed Einstein that "Henry Ford once told me that he, too, did not carve out his own life, but that all his actions were determined by an inner voice." Einstein seemed unsurprised. "Ford may call it his Inner Voice," he replied. "Socrates referred to it as his *daimon*. We moderns prefer to speak of our glands of internal secretion. Each explains in his own way the undeniable fact that the human will is not free.... I am not a psychologist, but it seems to me fairly evident that physiological factors, especially our endocrines, control our destiny."[22]

The question of free will is an ancient conundrum, one of the thorniest controversies in the history of human thought. For theologians, of course, it's phrased a little differently. The question is how to reconcile an omnipotent God, who created us the way we are (weak, sinful, and selfish), with His insistence that *we* bear the responsibility for making the right moral decisions. But the basic problem is the same: how do we harmonize a higher imperative to live an ethical life with omnipotent cosmic forces that don't really seem to allow for any agency on our part? Even one of the greatest scientists of all time had to grapple with this age-old problem, and we might be forgiven for hoping that maybe, just maybe, a thinker as original as Einstein was finally able to reach some kind of resolution.

Prepare to be disappointed. Einstein's halfhearted efforts to reconcile his strict determinism with his stern demands for ethical action reveal him at his least convincing. In one attempt to avoid the problem, he declared that despite being dedicated to determinism in principle, he was prepared to be pragmatic in practice. "If I wish to live in a civilized community," he conceded, "I must act as if man is a responsible being. I know that philosophically a murderer is not responsible for his crime, nevertheless I must protect myself from unpleasant contacts. I may consider him guiltless. But I prefer not to take tea with him."[23] This didn't apply only to murderers, of course; Einstein was talking about all of us. "In a sense, we can hold no one responsible," he continued. "I am a determinist. As such, I do not believe in Free Will." Whereas most people believed, at least in theory, that "man shapes his own life," Einstein insisted that "I reject that doctrine philosophically."[24] Caught between inscrutable inner compulsions and inexorable external causes, the conscious mind was mostly just along for the ride.

But at other times, Einstein seemed to say the exact opposite. He once claimed that "external compulsion can, to a certain extent, reduce but never cancel the responsibility of the individual."[25] On another occasion, he optimistically declared that "destiny will always be decided by what the individual feels, wills and does."[26] This has led one frustrated Einstein scholar to say that "his statements concerning answerability for one's actions were blatantly contradictory."[27]

Einstein's contemporaries agreed. He received numerous letters over the years complaining about his obvious inconsistency, and his replies leave much to be desired. In answer to one letter, he simply sidestepped the issue. "You make the statement that a consequent determinist opinion is really antiethical," he responded, because if there really were "freedom of the will," it would obviously imply "an incompleteness of the causal connection." But Einstein dodged this objection by arguing that "we should not make the fight for our ethical beliefs dependent on this scientific subtlety."[28] In another reply, he tried to be more reassuring. "You are worried by the conflict between the purely causal attitude of Spinoza and the attitude which is directed to active effort in the service of social justice," he wrote. But he explained that "I don't think that there exists a real conflict because our spiritual tensions, not only those of the passions but also those of the urges to achieve a more just order of society, belong to the factors which, with all others, partake of causality. It doesn't represent an inconsistency if we connect those spiritual states with the idea of purpose and goal."[29]

To put it mildly, these arguments are unsatisfying; in fact, they border on the incoherent (I can't make much sense of them, anyway). And Einstein seems to have sensed the inadequacy of his ideas on this all-important issue, because later in life he backtracked on his earlier beliefs. Whereas he'd once defended the idea that everything was determined, "for the insect as well as for the star," he later came to the conclusion that, in the case of human beings, our incredible complexity conferred on us a kind of conscious freedom lacking in other life-forms.[30] "While the whole life process of ants and bees is fixed down to the smallest detail by rigid, hereditary instincts," Einstein pointed out, "the social pattern and interrelationships of human beings are very variable and susceptible to change. Memory, the capacity to make new [mental] combinations, the gift of oral communication have made possible developments among human beings which are not dictated by biological necessities."[31] And so he was willing to admit that, "in a certain sense, man can influence his life through his own conduct, and that in this process conscious thinking and wanting can play a part."[32]

Not that this was the norm, of course. Einstein knew that "this inward freedom" was "an infrequent gift of nature."[33] And although he was convinced that there were "vast reserves of intelligence and courage which lie untapped within every human being," he also felt that "man, like every other animal, is passive by nature. Unless goaded by circumstance, he scarcely takes the trouble to reflect upon his condition and tends to behave as mechanically as an automaton."[34] In Einstein's opinion, it required exceptional spiritual strength to gain anything approximating conscious control over our own meandering course through the cosmos. "Only if outward and inner freedom are constantly and consciously pursued is there a possibility of spiritual development," Einstein decided, "and thus of improving man's outward and inner life."[35]

But Einstein believed that with "a good deal of conscious thought and of self-education," we *could* exert some control over our own destiny.[36] We could attain a "freedom of the spirit," an "independence of thought from the restrictions of authoritarian and social prejudices as well as from unphilosophical routinizing and habit in general."[37] And since we were hardly the first people to grapple with these dilemmas, we could look to the past for encouragement and inspiration. As usual, Einstein saw the most admirable exemplars in Greece and the East. It was "undeniable," he thought, "that the enlightened Greeks and the old Oriental sages had achieved a higher level in this all-important field."[38] In particular, he felt that "the old Chinese sages understood and proclaimed that the most important factor in giving shape to our human existence is the setting up and establishment of a goal; the goal being a community of free and happy human beings who by constant inward endeavor strive to liberate themselves from the inheritance of anti-social and destructive instincts."[39] In other words, we *were* capable of attaining a certain degree of freedom, but this autonomy had to be achieved through intensive inner work. Agency had to be *earned*.

This is not a new idea. In a wide variety of guises, it's actually one of the central messages of most spiritual traditions. Almost every religion agrees that, for the most part, people are prisoners—of fate, of karma, of the devil—and that only the inner struggle of the earnest spiritual practitioner can confer any real autonomy. But for Einstein, the struggle for freedom was no longer a task only for exalted saints and enlightened sages. "Everyone has been given an endowment," he thought, "that he must strive to develop in the service of mankind. This cannot be brought to completion through the threat of a God who will punish man for sin, but only by challenging the best in human nature."[40] And since no supernatural power would ever summon us to such a challenge, it was up to us to bring out the best in ourselves—and each other. "Mankind must exalt itself," Einstein insisted. "*Sursum corda* [Lift up your hearts!] is always its cry."[41]

Still, Einstein never really spelled out how this was possible—how, precisely, a little part of Infinity could attain some degree of freedom in a supposedly deterministic universe. But the Einstein scholar Robert Goldman offered an intriguing explanation that unravels the enigma. "In a world of being, consisting of a complex network of spacetime interrelations, the creative individual can be looked upon as an essential node which has helped determine the pattern of interrelation," Goldman explained.[42] "Since each part of the great pattern of being in spacetime is dependent on the parts to which it relates, the part under consideration itself...plays more than a passive role in the determination of the nature of its own being: it helps determine the other parts which in turn determine itself."[43] These considerations led Goldman to conclude that even Einstein's ironclad determinism "does *not* imply fatalism. It does *not* imply that the individual has no effect on the passage of events."[44] More than two thousand years ago, the Pythagoreans had already reached a similar perspective on the apparent paradox of free will. "Because each part is linked to the whole through *harmonia*," they held, "every action has its repercussions, either beneficial or not, for which the individual is supremely responsible."[45]

Despite his extremist stance on determinism, Einstein ultimately agreed. Although we couldn't claim to be free in any fundamental sense, we still exerted a force in the world, like the feeble pull of our own little planet on the sun. So no matter how small the role, every person had a part to play. "Each of us has to do his little part," Einstein insisted, "toward transforming the spirit of the times."[46] Our inextricable ties to all things everywhere meant we inevitably impacted the world around us; our actions rippled and reverberated through all of reality. "Human beings are *not* condemned," as Einstein concluded near the end of his life, "to be at the mercy of a cruel, self-inflicted fate."[47] Even the tiny nexus of the individual had an impact on the overall network. "What we do," Einstein decided, "is of supreme importance to all humanity, to history, to human destiny."[48]

But what should we do? There were no easy answers, and Einstein was the first to admit just how extraordinary it was that existence didn't come with any ethical instructions. "Strange is our situation here upon earth" was the opening line of his personal manifesto. "Each of us comes for a short visit, not knowing why, yet sometimes seeming to divine a purpose."[49] Conscious beings had suddenly appeared on the cosmic scene, and we found ourselves inhabiting a mysterious world without the faintest idea why we were here, how we should behave, or where we were going. Every religion in the world offers prepackaged answers to these perennially perplexing questions. But Einstein thought it was high time we forgot about the fairy tales of our forebears and fearlessly faced up to the enigma of existence. "Life is a mystery to be lived," he said, "not a problem to be solved."[50]

And it was a mystery we had to face on our own. "Ethics," Einstein insisted, was "an exclusively human concern with no superhuman authority behind it."[51] With this seemingly inoffensive opinion, Einstein was actually going up against an explicit assertion of almost every great faith across the ages. For our received religions, the real rules of morality had been laid down long ago by an invisible God, and it was precisely the superhuman source of moral laws that made them *laws*. Human beings had neither the right to question the correctness of these commandments, nor the requisite wisdom to try to reform them. All that was expected of us was obedience to these eternal ethical ideals.

Einstein begged to differ. "Ethical behavior should be based effectively on sympathy, education, and social relationships," he said. "No religious basis is necessary."[52] By no means did he believe we should abandon all moral standards. He acknowledged that "we need concepts to order our own existence—for example, this is good and that is evil."[53] But he worried that "these concepts have been accepted for so long that no one questions their authority. We forget their human origin and think they represent indestructible truth."[54] It was essential to remember that even our oldest ethical ideals were human inventions. "The existence and validity of human rights is not written in the stars," Einstein reminded us. "It was enlightened men who, in the course of history, conceived and taught the ideals concerning the conduct of human beings toward one another."[55] The idea that we should blindly obey the ethical axioms we'd inherited from our ancestors was becoming an insult to our intelligence, "incompatible," as Einstein put it, "with the integrity of a modern cultured person."[56]

Not that Einstein wanted to remove religious feelings from the ethical realm altogether. Far from it. Although he was convinced that "the conquest of prejudice and superstition" was a critical part of our progress as a species, he also contended that "the clearing away of obstacles does not by itself lead to an ennoblement of social and individual life."[57] Because "along with this negative result," he realized, "a positive aspiration and effort for an ethical moral configuration of our common life is of overriding importance."[58]

It was up to us to devise this positive program ourselves. True, any ethical axioms we opted to embrace would ultimately be arbitrary human inventions. But they would be *ours*: ideals created by the humble inhabitants of Earth and aligned with our own highest aspirations. All of us were involved in the alchemical quest to take the dross of our old religious rules and forge a new ethical alloy for the future, and Einstein was unapologetically optimistic about the prospects of this adventure. He believed that "there lies before us, if we choose, continued progress in happiness, knowledge, and wisdom."[59]

And yet Einstein always made it clear that "my God is too universal to concern himself with the intentions of every human being."[60] So why bother to be good, if there were no payoffs in Heaven or punishments in Hell? Why not simply serve our own selfish interests and do as we please? For Einstein, even asking this question was contemptible. With a rare rhetorical flourish, he once wrote: "If someone asks, 'For what purpose should we help one another, make life easier for each other, make beautiful music together, have inspired thoughts?' he would have to be told, 'If you don't feel the reasons, no one can explain them to you.' Without this primary feeling we are nothing and had better not live at all."[61] In other words, elevated ethical action could only be for its own sake; we owed it to *ourselves* to act in accordance with the highest possible principles. As far as Einstein was concerned, anything less would be beneath us. "What a miserable creature man would be," he thought, "if he were good not for the sake of being good, but because religion told him that he would get a reward after this life, and that if he weren't good he'd be punished."[62]

But where would we even get our idea of what was *good*? Where would our ethical principles come from, if not religious scriptures and the revelations of saints? "The world needs new moral impulses, which, I'm afraid, won't come from the churches," Einstein said.[63] "Perhaps," he suggested, "those impulses must come from scientists."[64] It sounds almost absurd to suggest that science could ever serve as the source of our ethics, but Einstein was dead serious. If we were going to build an enduring ethical system all people could adhere to, we couldn't keep relying on religious fairy tales and we couldn't keep ignoring physical reality. The old religious monopoly on morality could no longer be maintained.

"Needless to say," Einstein noted, "one is glad that religion strives to work for the realization of the moral principle." Nonetheless, he made it clear that "the moral imperative is not a matter for church and religion alone."[65] What was needed now was neither a list of new commandments nor a litany of new vows, but rather "the deliberate nurturing of the moral sense also outside the religious sphere."[66] If we could find a way to amalgamate the insights derived from scientific experiment and the intuitions developed from spiritual experience, then maybe a workable global ethics would be within reach. "Believe me," Einstein said, "what the religious leaders were never able to do—to tear down the barriers between races and religions—the scientists of our age may do."[67]

Progress of this kind would require an innovative integration of science and spirituality, and Einstein knew that such a synthesis was still only a fantasy. The old antagonism between science and religion was alive and well, and Einstein readily acknowledged that religion alone wasn't to blame. Although he felt that most religions suffocated the human mind, he also admitted that, so far, science too had done more to stifle the human spirit than to stimulate it.[68] Science *seemed* to insist on a materialist mindset, *seemed* to suggest that Nature was all mindless mechanism. And Einstein felt forced to admit that "the causal and objective mode of thinking—though not necessarily in contradiction with the religious sphere—leaves in most people little room for a deepening religious sense."[69] So it was unsurprising that the increased secularization of society had coincided with "a serious weakening of moral thought and sentiment."[70]

Such a weakening was of course exactly what religious traditionalists had always warned of and worried about. Orthodox believers had long argued that abandoning our belief in God-given ethical maxims would lead to lawlessness. With only the ever-shifting sands of scientific knowledge serving as foundation, the edifice of ethics was sure to crumble. The attempt to transcend the tried-and-true ethical ideals of past religious teachers was the epitome of arrogance, and it wasn't hard to see that humanity had already paid a high price for its hubris. For all our pretensions of progress, it was present-day people who'd committed the most contemptible crimes in history. The horrors of the twentieth century seemed to provide unequivocal proof that secularization and scientific progress alone were inadequate means of building a better world.

In the eyes of some people, Einstein himself epitomized the problem. Wasn't it his equation $E = mc^2$ that had shown the equivalence of matter and energy and led to the disastrous development of the atomic bomb, unleashing an unprecedented destructive force on the world? Didn't his theory of relativity imply that all things were relative even in the once-reliable realm of physics, removing the last speck of solid ground on which the human mind could stand? These concerns might sound quaint today, but in Einstein's time many people worried that his theories would further erode the foundations of mainstream religion, leaving behind only the wreckage of moral relativism. Even eminent religious authorities were alarmed at the imagined implications. The archbishop of Boston begged his followers not to read anything about relativity and berated Einstein for his "befogged speculation," which he accused of "producing universal doubt about God and his Creation."[71] And the archbishop of Canterbury, concerned that the new physics would corrode good Christian values, also wanted to know "what effect relativity would have on religion." Einstein was unequivocal. "None," he replied. "Relativity is a purely scientific matter and has nothing to do with religion."[72]

And yet Einstein's resolute insistence that relativity had nothing to do with moral relativism has mostly been ignored. His physics continues to be misunderstood as a source of loose morals and postmodern sensibilities, and it's not just religious leaders who are guilty of this gaffe.[73] In its final issue of the millennium, in December 1999, *Time* magazine proclaimed Einstein the "Person of the Century" and didn't miss the opportunity to perpetuate this myth. "Relativity paved the way for a new relativism in morality, art, and politics," claimed the cover article. "There was less faith in absolutes, not only of time and space but also of truth and morality."[74]

But nothing could have been further from Einstein's mind than moral relativism. In fact, he advocated the exact opposite: a "universal moral attitude" that could be adopted by all.[75] His hope was that humankind could be united by a simple set of secular values that were nonetheless suffused with a spiritual vitality. All people "should let their conduct be guided by the same principles," he suggested, "and those principles should be such, that by following them there should accrue to all as great a measure as possible of security and satisfaction, and as small a measure as possible of suffering."[76] True to form, Einstein envisioned a parsimonious ethical system that emulated the precision of mathematics. "If we can agree on some fundamental ethical propositions," he proposed, "then other ethical propositions can be derived from them, provided that the original premises are stated with sufficient precision. Such ethical premises play a similar role in ethics, to that played by axioms in mathematics."[77]

In no way was Einstein endorsing the naïve notion that we needed to derive our ethical axioms from scientific *knowledge*. It was clear to him that "the content of scientific theory itself offers no moral foundation for the personal conduct of life," and he even proclaimed himself "perfectly convinced" that "every attempt to reduce ethics to scientific formulas must fail."[78] He never claimed that science, as such, could transcend the traditional teachers of humankind. "Men like Confucius, Buddha, Jesus, and Gandhi," Einstein said, "have done more for humanity with respect to the development of ethical behavior than science could ever accomplish."[79] But "on the other hand," he pointed out, "it is undoubtedly true that scientific study of the higher kinds and general interest in scientific theory have great value in leading men toward a worthier valuation of the things of the spirit."[80] Because for Einstein, there was much more to science than the mere accumulation of facts. There was also the spirit that animated the scientific endeavor: the yearning for unity and the fervent faith that it could be found. And in this more sublime sense, he felt that "scientific theory brings into play the higher spiritual faculties, and anything that does so must be of high importance in the moral betterment of humanity."[81]

For many theologians in Einstein's time, and many people today, abandoning the absolute ethical injunctions of the old religions meant being set adrift in a sea of moral uncertainty for all eternity. But Einstein didn't share this despair. Instead, he opted to anchor his ethical ideals in his scientific intuition that all things were interconnected. "Scientists in the tradition of Galileo, Kepler, and Newton," Einstein noted, "devoted their lives to proving that the universe is a single entity, in which, I believe, a humanized God has no place."[82] And if all things actually *were* only interdependent aspects of a "single entity," Einstein thought, we ought to act accordingly. Once we accepted the simple axiom of the nonduality of

Nature, then the natural next step was to accept nonviolence as the very essence of ethics.

Einstein didn't expect anyone to simply take him at his word. On the contrary, he maintained that this mind-expanding insight was an essential experience for all future generations. "O Youth: Do you know that yours is not the first generation to yearn for a life full of beauty and freedom?" he once wrote. "Do you know that all your ancestors felt as you do—and fell victim to trouble and hatred? Do you know, also, that your fervent wishes can only find fulfillment if you succeed in attaining a love and an understanding of people, and animals, and plants, and stars, so that every joy becomes your joy and every pain your pain? Open your eyes, your heart, your hands," he urged us. "Then will all the earth be your fatherland, and all your work and effort spread forth blessings."[83]

In other words, only immediate individual experience of an extraordinary kind could convince us that in essence we were all akin, all equally aspects of the Infinite. "If man becomes more aware of his dignity as a cosmic being than of his ego in the flesh," Einstein thought, "our world would then have peace."[84] And so all of us had a personal responsibility to strive to see we were part of Infinity.

8

PART OF INFINITY

The soul given to each of us is moved by the same living spirit that moves the universe.

—Albert Einstein, in conversation with William Hermanns

To SEE ETERNITY IN EVERY ENTITY WAS THE HARDEST TASK AND THE highest goal of Einstein's spirituality. This ambition seemed avant-garde to his contemporaries and abhorrent to his critics, but actually Einstein was neither innovator nor heretic. He was merely the latest link in a long chain, the living embodiment of an ancient legacy that held that the human mind was at home in the universe and identical to the Infinite. Passed through the limiting prism of human perception, the Infinite appeared finite and the One became many. But Einstein insisted that, in the end, "individuality" was "an illusion."[1] A person's apparent independence was "a kind of optical delusion of his consciousness," and for Einstein, "the striving to free oneself from this delusion" was "the one issue of true religion."[2]

Dispelling the delusion didn't require us to retreat to a remote monastery or sequester ourselves on a secluded mountaintop. Seen through awakened eyes, all things offered the opportunity to experience Infinity: the baffling beginning of every birth; the enigmatic departure of every death; the fragile little layer of life blooming from the Earth's surface; even the endless expanses of empty space separating the stars. The Infinite danced all around us. Every instant offered another chance for enchantment.

And yet for Einstein, the ultimate aim was not simply the enlightenment of the individual. Perceiving oneself as a part of Infinity implied more than just leaving all personal cares behind. It meant being on a mission to destroy the dichotomy between microcosm and macrocosm. It meant seeing and believing, deep in one's bones, that we'd all been born from the same source billions of years ago. And at its most basic level, it meant becoming someone who could commune with the cosmos and create something that could serve the whole human community.

An intimacy with Infinity informed every aspect of Einstein's existence. In every sphere he sought an all-encompassing unity, and in every domain he felt the inadequacy of our efforts so far. Science was still simplistic; ethics was still exclusionary; religion still needed to be re-created and revitalized.

For Einstein, making advances on all these fronts was essential. But beyond that, he also sensed an even deeper deficiency: our self-concept, our very sense of our own souls, was severely limited. "The fact that man produces a concept 'I' besides the totality of his mental and emotional experiences or perceptions does not prove that there must be any specific existence behind such a concept," Einstein thought. "We are succumbing to illusions produced by our self-created language, without reaching a better understanding of anything."[3] In Einstein's opinion, we were ensnared by narrow notions of our own existence, and this ignorance of our essential nature inevitably entailed anguish. "All human beings, whatever their position in society, are suffering from this," Einstein felt. "Unknowingly prisoners of their own egotism, they feel insecure, lonely, and deprived of the naïve, simple, and unsophisticated enjoyment of life."[4]

Overcoming these constraints was the very core of the cosmic religion. "The main task of the spirit," as Einstein saw it, was "to free man from his ego."[5] What was needed was a modification of the mind, a modulation of the mental filters that kept us confined to this minimalist model of the self. And in a sense, this was an ancient quest, the oldest adventure of the human spirit. "Every cultural striving," Einstein thought, "whether it be religious or scientific, touches the core of the inner psyche and aims at freedom from the Ego."[6]

Beyond the prison of the personal was a purer place that provided a superior perspective. Liberated from our usual limitations, we could see the links connecting us with all things, from the littlest life-form to the luminous galaxies light-years away. The central insight was easy to express. "I am a part of infinity," Einstein exclaimed. "I see everything in *specie aeternitatis*."[7] The statement itself, and even the way Einstein worded it, is pure Spinoza. Three centuries before Einstein, Spinoza had similarly declared that "the mind is eternal insofar as it conceives things under a form of eternity."[8]

Einstein wasn't the first thinker to follow Spinoza's example and embrace his own immanent Infinity. Einstein's other favorite philosopher, Schopenhauer, also focused on this same fascinating phrase. "Raised up by the power of the mind, we relinquish the ordinary way of considering things," Schopenhauer said. "We forget our individuality." Schopenhauer was certain that "it was this that was in Spinoza's mind when he wrote: 'The mind is eternal in so far as it conceives things from the standpoint of eternity.'"[9] And for Schopenhauer, Spinoza's style of self-transcendence was equivalent to the enlightenment of the East. In words Einstein probably knew well, Schopenhauer explained how the rare mind that had "found itself again in everything" experienced "reabsorption in *Brahman*, or the *Nirvana* of the Buddhists."[10] There's no doubt Einstein had similar experiences. "I feel myself so much a part of everything living," he once said, "that I am not in the least concerned with the beginning or ending of the concrete existence of any one person in this eternal flow."[11]

Still, even an immediate experience of the interdependence of all things wasn't enough. In order to fully feel one's affinity with Infinity, time had to be transcended. Einstein once declared that "the distinction between past, present, and future is only a stubbornly persistent illusion."[12] He also spoke of certain sacred instants when time stood still and the eternal became apparent everywhere. "There are moments when one feels free from one's own identification with human limitations and inadequacies," he wrote. "At such moments, one imagines that one stands on some spot of a small planet, gazing in amazement at the cold yet profoundly moving beauty of the eternal, the unfathomable: life and death flow into one, and there is neither evolution nor destiny; only being."[13]

These magical moments where Einstein felt himself "dissolved or merged into Nature" became more and more frequent as his life neared its end.[14] "The strange thing about growing old," he said, "is that the intimate identification with the here and now is slowly lost; one feels transposed into infinity, more or less alone, no longer in hope or fear, only observing."[15] He'd learned the ultimate lesson of his favorite philosophers well: time was a trick of the mind, a pure form of perception that kept us blind to the boundless.

And after transcending time and space, Einstein saw what countless sages had seen across the ages. "Body and soul are not two different things," he boldly declared, "but only two different ways of perceiving the same thing."[16] Philosophically speaking, there was no real distinction between physical science and self-reflection. "Physics and psychology," as Einstein put it, "are only different attempts to link our experiences together by way of systematic thought."[17] There's no mistaking Einstein's meaning: he was arguing for what philosophers call *metaphysical monism*—a doctrine of nonduality that denies the dichotomy of mind and matter. And as with so much of Einstein's spirituality, Spinoza proves to be a principal source. "To Spinoza," Einstein explained, "the spiritual and the material" were "merely different manifestations of one universal, deterministic reality."[18] And *this* was why Einstein saw Spinoza as "the greatest of modern philosophers"— because he was "the first philosopher who deal[t] with the soul and the body as one, not as two separate things."[19]

Einstein is right: Spinoza deserves much of the credit for reviving monism in modern times. But as Einstein well knew, Spinoza's notion of nondualism was nothing new. Many mystics over the millennia had maintained that mind and matter, microcosm and macrocosm, were ultimately equivalent. The parallels are particularly obvious among Einstein's favorite thinkers, and no doubt he noticed echoes of this feeling throughout the world's spiritual systems. The English translation of the Chandogya Upanishad in his personal library declared that "the Infinite indeed is below, above, behind, before, right and left—it is indeed all this." And since the Upanishadic sages saw "the Infinite as the I," the same could also be said of oneself: "I am below, I am above, I am behind, before, right and left—I am all this."[20]

Taoism also defended this deep affinity between the part and the whole. "Tao is the universal principle, Te...is the part which the individual has in Tao," Richard Wilhelm commented in Einstein's copy of the Tao Te Ching.[21] And although Einstein might not have known it, even the Pythagoreans had agreed we were part of Infinity. "Pythagoras said that man was a *microcosm*," the ancient writer Photius tells us, "because he contains all the powers of the cosmos."[22]

Perhaps no one expressed this insight more poetically than Schopenhauer. "If we lose ourselves in contemplation of the infinite greatness of the universe in space and time, meditate on the past millennia and on those to come...we feel ourselves reduced to nothing," he wrote. But once we realized that we were "one with the world," then we would be "not oppressed but exalted by its immensity." Schopenhauer considered his own experience of the Infinite as merely a modern variation on a venerable tradition. "It is the felt consciousness," he concluded, "of what the Upanishads...express repeatedly in so many different ways."[23] And similarly, in proclaiming that he was a part of Infinity, Einstein was only giving his own idiosyncratic expression of an ancient ideal, the oldest insight in the world.

But for Einstein, transcending the ego was much more than an uplifting subjective experience—it was an urgent moral imperative. Since a personal experience of nonduality was the best promoter of nonviolence, he believed that "liberation from the bondage of the self constitutes the only way towards a more satisfactory human society."[24] Peering beneath the surface play of personality was therefore the noblest possible endeavor, the real test of any mind. "The true value of a human being," Einstein declared, "is determined primarily by the measure and the sense in which he has attained to liberation from the self."[25]

Einstein didn't claim he'd attained this aim; he never insinuated he was in any way enlightened. His own moments of self-transcendence were always fleeting, and actually he assumed that "nobody is able to achieve this completely."[26] But the unattainability of the ultimate goal didn't undermine its importance at all. "The striving for such achievement," Einstein said, "is in itself a part of the liberation."[27] Ultimately, it wasn't some perfect attainment of enlightenment that was essential, but rather untiring efforts in service of this sacred end. "A person who is religiously enlightened," Einstein said, "appears to me to be one who has, to the best of his ability, liberated himself from the fetters of his selfish desires."[28]

For all his flaws, the aura of the Infinite around Einstein was obvious to those close to him. "When one was with him on the sailboat, you felt him as an element," recalled his stepdaughter Margot. "He had something so natural and strong in him because he was himself a piece of nature."[29] One biographer who knew him well thought that "Einstein, with his feeling of humility, awe, and wonder and his sense of oneness with the universe, belongs with the great religious mystics."[30] And Rabindranath Tagore felt that Einstein's "transcendental materialism" transported him to "the frontiers of metaphysics, where there can be utter detachment from the entangling world of self."[31]

But although many people appreciated Einstein's otherworldly air, even close friends were curious about the origin of his emancipation, however imperfect, from the ego. Hedwig Born, wife of the Nobel Prize–winning physicist Max Born, kept up a lively correspondence with Einstein and once sent him some sonnets she'd composed on the Indian ideal of liberation. "I try to express the thought that love alone binds and liberates one (from the ego) at one and the same time," she told Einstein in a letter, explaining how her poems "came into being from [the] Indian [notion of] 'non-attachment.'" And she concluded her letter with a curious question. "It seems to me that you have achieved such non-attachment," she noticed, "but how?"[32]

We ought to ask ourselves the same question. Although it's obvious that Einstein had spiritual experiences of some sort or another, they seem to have stemmed from an unusual source. We tend to think of transcendent experiences as the outcome of intensive spiritual work in utter seclusion from the world: Buddha under the bodhi tree, Jesus in the desert, Mohammed in the cave on Mount Hira. And certainly Einstein always sought the single most essential element of the spiritual life, insisting that "I always loved solitude."[33] At times he even sounds like a cranky old sage trying to escape the cacophony of so-called civilization. "How conducive to thinking and working the long sea voyage is," he proclaimed on his way back from visiting Asia in 1923.[34] The peace and quiet must have reminded him of the Buddhist monasteries he'd recently visited, because he explicitly compared his seclusion to the cloistered life. "Such a sea voyage," he said, "is a splendid existence for a ponderer—it's like a monastery."[35] He even surmised that "perhaps, someday, solitude will come to be properly recognized and appreciated as the teacher of personality. The Orientals," he observed, "have long known this."[36] But even though Einstein was well aware of Eastern meditation practices and admitted to being "very impressed" by the "serenity and selflessness" of monks he'd seen meditating in Asia, he never performed any spiritual practices himself, and he always insisted that "I don't care to be called a mystic."[37]

Still, it's not as if sitting quietly with legs crossed is a prerequisite of all profound experience. In fact, in ancient India, a purely (or at least primarily) intellectual path to Oneness was recognized thousands of years ago. Some of the oldest sacred scriptures of Hinduism speak of a path for people with a strong intellectual inclination—deep thinkers for whom blind faith and fawning devotion are unsuitable avenues to the absolute.

This was the road known as the *jñāna mārga*, the Path of Knowledge.[38]* *Jñāna* was a radical means of experiencing the Infinite through intellectual insight, a mystical method "unique in the history of the world," in the words of one scholar.[39]

One of the most adamant advocates of this intellectual path was the Indian philosopher Shankara (c. 700–750 C.E.), long acknowledged as the ultimate exponent of the nondualist Advaita Vedanta school. Shankara believed that the ultimate reality, Brahman, was immanent everywhere; it was "everything that can be experienced in this universe."[40] Only our self-limiting beliefs blinded us to the fact that Brahman was right before our eyes. "Because of the ignorance of our human minds," he explained, "the universe seems to be composed of diverse forms." But Shankara maintained that all multiplicity was only an illusion; in actuality, everything was "Brahman alone."[41] And much like Einstein, Shankara insisted that the origin of this illusion was "the ego, the first-begotten child of ignorance."[42]

The fundamental aim of Shankara's philosophy was to find a way "to be free from the fetters forged by ignorance."[43] But when it came to "the realization of one's true nature," he believed the traditional religious rites were worthless.[44] "Men may recite the scriptures and sacrifice to the holy spirits," he said, "they may perform rituals and worship deities—but, until a man wakes to knowledge of his identity with the Atman, liberation can never be obtained."[45] For Shankara, only a deep intellectual insight could free us from our false sense of individuality. "Nothing but the sharp sword of knowledge," he said, "can cut through this bondage."[46] And so he didn't ask for faith or devotion from his followers. Instead, he insisted that the ideal disciple "should be intelligent and learned, with great powers of comprehension, and able to overcome doubts by the exercise of his reason."[47] By following this rational path to realization, Shankara promised, "the wise man reaches that highest state, in which consciousness of subject and object is dissolved away and the infinite

---

* Pronounced kind of like "nyaa-na," the Sanskrit word *jñāna* is cognate to the modern English *know* and the ancient Greek *gnosis*. The words are linked through the old Indo-European language that gave rise to them all.

unitary consciousness alone remains—and he knows the bliss of Nirvana while still living on earth."[48]

Einstein probably had at least a passing familiarity with Shankara, because Schopenhauer refers to him several times when discussing non-dualism.[49] And as it turns out, Schopenhauer was searching for essentially the same sort of enlightenment: "liberation," as he put it, "from the one-sidedness of an individuality which does not constitute the innermost kernel of our true being."[50] Even though Schopenhauer thought that the "sublime authors of the Upanishads" who'd inspired Shankara could "scarcely be conceived as mere human beings," and attributed their deep insight to an "immediate illumination of their mind," he also held that rational realization could carry us just as far. "Thorough reflection," he thought, "as carried through by Kant's great mind, also leads to just the same result by a different path."[51]

It's no secret that Spinoza harbored similar aspirations. And in another uncanny parallel with Eastern ideas, he somehow arrived at his own understanding of an intellectual path to the Infinite. In his *Ethics*, he spoke of a "third kind of knowledge," different from knowledge gained through the senses or from reasoning—an enigmatic "intuition" that gave us intimate knowledge of the Infinite.[52] "The highest virtue of the mind," he maintained, "is to understand things by this third kind of knowledge,"[53] because "from the third kind of knowledge there necessarily arises the intellectual love of God."[54] This *amor Dei intellectualis* was not the normal kind of love, but rather "love toward something immutable and eternal," a rational reverence for the ultimately real.[55] And through this third kind of knowledge, Spinoza believed, "the mind can bring it about that all the affections of the body—i.e., images of things—be related to God."[56] In other words, we could realize that all our various sensory experiences were really only surface manifestations of the one underlying Substance "common to all things."[57] By bringing all our mental powers to bear, we could "perceive things in the light of eternity."[58]

With dedicated effort, the insight afforded by this *amor Dei* would become ever deeper and more enduring. Spinoza even described for us how his own experience of the divine developed over time. "Although at first these intermissions were rare and of very brief duration," he admitted, "nevertheless, as the true good became more and more discernible to me, these intermissions became more frequent and longer."[59] The ultimate goal was for this glorious intuition to "continue to grow more and more and engage the greatest part of the mind and pervade it" until we reached "the supreme good"—which was nothing other than "knowledge of the union which the mind has with the whole of Nature."[60] And since Spinoza, like Shankara before him, asserted that all was One, we were really only worshipping ourselves. "The mind's intellectual love toward God is part of the finite love wherewith God loves himself," Spinoza explained, "not insofar as he is infinite, but insofar as he can be explicated through the essence of the human mind."[61] And so the *amor Dei intellectualis* was not the creature worshipping its Creator, but only the Infinite encountering the Infinite, the divine adoring the divine.

Einstein was well aware of Spinoza's remarkable system of rational spirituality, referring to it as "that mysterious inner consecration which Spinoza so often emphasized under the name *amor intellectualis*."[62] But for Einstein, this was more than just a rational kind of religious feeling— it was the real source of the radical progress science had made in revealing the secrets of Nature. "If those searching for knowledge had not been inspired by Spinoza's *Amor Dei Intellectualis*," he insisted, "they would hardly have been capable of that untiring devotion which alone enables man to attain his greatest achievements."[63] For Einstein, this intellectual love for the absolute was the inexhaustible source that sustained the scientific seeker in the solitary search for knowledge.

And even as this rational approach to the divine revealed the roots of the natural world, it also expanded our narrow notions of ourselves. In Einstein's eyes, the "striving" for "rational unification" of the physical world was a sort of spiritual discipline that could transform our very sense of self. "Whoever has undergone the intense experience of successful advances [in science] is moved by profound reverence for the rationality made manifest in existence," he explained. "By way of the understanding he achieves a far-reaching emancipation from the shackles of personal hopes and desires."[64] In other words, intellectual insight led to ego dissolution; the pursuit of science, practiced in a pure way, promoted self-transcendence.

To Einstein, it wasn't the *attainment* of knowledge that mattered so much as the struggle itself. "Measured objectively, what a man can wrest from Truth by passionate striving is utterly infinitesimal," Einstein admitted. "But the striving frees us from the bonds of the self."[65] And so Einstein insisted that "if it is one of the goals of religion to liberate mankind as far as possible from the bondage of egocentric cravings, desires, and fears, scientific reasoning can aid religion."[66] Naturally, this is not what most religious people have in mind when they think of knowing the numinous. But Einstein saw no inconsistency. "This attitude appears to me to be religious," he said, "in the highest sense of the word. And so it seems to me that science not only purifies the religious impulse of the dross of its anthropomorphism but also contributes to a religious spiritualization of our understanding of life."[67]

There's a strong tendency today to downplay the religious element in scientific inquiry. It seems almost absurd to think of scientists as spiritual people, and downright blasphemous to see them as beings on the path to enlightenment. But what Einstein understood was that really exceptional insights didn't inflate the ego: they emancipated us from individuality by aligning us with all of existence. There have always been plenty of egotistical scientists, of course, and Einstein was well aware of the personal faults and failings of even his highest scientific heroes, like Newton and Galileo.[68] But Einstein believed that in their best moments, at least, all truly great minds were bewildered by the beauty of their own inspirations. Reason could barely begin to comprehend its own insights into the workings of reality, or what the efficacy of our understanding actually implied about the relationship between consciousness and cosmos. The stupefied ego struggled to understand how the microcosm of the human mind could make sense of the macrocosm that is the material universe.

And the greater the discovery, the greater the debt these thinkers felt they owed to the divine. It was in this sense that James Clerk Maxwell spoke of the pursuit of science as a spiritual quest, "an embodiment of the work of Eternity."[69] In a remarkable parallel with Einstein's own words,

Maxwell even insisted that anyone who succeeded in revealing Nature's secrets became "a partaker of Infinity."[70] And so Einstein was able to find inspiring examples of ego transcendence not only among the most eminent philosophers of all times, but also among the ingenious physicists who were the forerunners of his own all-encompassing theories.

The idea that liberation comes from purifying knowledge rather than a pure heart is rare in religions today. In most of our existing traditions, especially the major monotheistic faiths, a combination of good behavior and God's grace has always been seen as the source of salvation. In contrast, Einstein embraced the subtle psychological insight put forth by Spinoza, Shankara, and so many other Eastern sages: that our seeming separateness was only an *epistemological error*, a misunderstanding created by our limited minds. "Actual experience takes place in kaleidoscopic particular situations," as Einstein put it. "The great variety of the external situations and the narrowness of the momentary content of consciousness bring about a sort of atomizing of the life of every human being."[71] Einstein was echoing Schopenhauer here, who'd made the same point a century before. "The eyes of the uncultured individual are clouded, as the Indians say, by the veil of Maya," Schopenhauer wrote. "In this form of his limited knowledge, he sees not the inner nature of things, which is one, but its phenomena as separated, detached, innumerable, very different, and indeed opposed."[72]

And yet what the mind had mistaken as many it could also merge again into One. There were different ways to reveal this deeper unity, of course. But for someone like Einstein, science was the surest spiritual path. "In a man of my type," he maintained, the deepest understanding could be gained when the mind "disengages itself to a far-reaching degree from the momentary and the merely personal and turns toward the striving for a conceptual grasp of things."[73]

For Einstein, a conceptual grasp meant not just a proliferation of facts or an accumulation of observations, but a comprehensive synthesis of all these separate strands of information. Rational knowledge meant a deep reckoning with the nature of reality and a nuanced appreciation of our true place in the transcendental tapestry. And since real science and real religion were indistinguishable for him, they inevitably shared the same goal: seeing unity in multiplicity, or what Einstein called "striving after the rational unification of the manifold."[74]

Einstein could hardly have been more explicit that disciplined rational exertion had led him directly to deep religious experience. "Through my pursuit of science," he said, "I have known cosmic religious feelings."[75] And so he saw further rational synthesis as the preeminent program not only of rigorous science, but also of real spirituality. "The further the spiritual evolution of mankind advances," he argued, "the more certain it seems to me that the path to genuine religiosity does not lie through the fear of life, and the fear of death, and blind faith, but through striving after rational knowledge."[76] Call it what we will—liberation, Heaven, enlightenment—for Einstein this was not a place one went or a reward one received. It was a fundamental reorientation of attention, a revolutionary alteration in perception that revealed a pure unity underlying all things, from the infinite phenomena of the physical world to the endless fluctuations of thought and feeling.

All things are actually One: it's an alluring idea with an ancient ancestry. From the Vedantic Brahman to the Pythagorean One to Spinoza's Substance, many schools of thought through the ages have insisted that some integral substratum is the ultimate origin of all apparent multiplicity. And yet this notion has always been abstract and unprovable—an inspiring metaphorical expression or an ineffable mystical experience, but never an actual model of existence. Always wildly ambitious, Einstein set himself an audacious task: to bring this ethereal philosophy down to earth and finally put it on a firm physical footing. He wanted to demonstrate that what he'd discovered through deep philosophical reflection was demonstrably true of physical reality.

It's easy to forget just what a radical program this is. Today, we tend to take it for granted that the same physical laws apply everywhere; that the same kinds of atoms occupy every corner of the cosmos; that, even in distant galaxies, other planets orbit other suns much like our own. We're so accustomed to this unified worldview that we rarely think about how recent, and how revolutionary, it really is. But this is not how our ancestors approached existence. From time immemorial, human beings had been looking up at the sky and the stars and assuming things must be fundamentally different in the world above. Up there was the domain of the gods, an unchanging and perfect realm with only the most tenuous of connections to our corrupt and crumbling world below.

The geniuses of the Scientific Revolution changed all that. In formulating physical laws that applied always and everywhere, the first great scientists began to collapse the immense distances and abolish the imagined differences between Heaven and Earth. Nowhere is this more apparent than in the work of Isaac Newton. The old story about an apple falling on his head might be a myth, but in a single simple image it sums up his most astonishing and important insight: the same force of gravity governs both human bodies on Earth and the heavenly bodies above. In a similar vein, Michael Faraday, when he united electricity and magnetism, showed that the light reaching us from distant stars was simply a rippling of the same electromagnetic force field that binds our body's biomolecules together.[77]

Einstein had immense esteem for both Newton and Faraday, but he saw James Clerk Maxwell as the greatest unifier and the real revolutionary. Following in the footsteps of his mentor Michael Faraday, Maxwell abandoned Newton's notion of material points moving through empty space and instead reconceptualized "physical reality as represented by continuous fields."[78] Einstein considered "this change in the conception of reality" to be "the most profound and fruitful one that has come to physics since Newton."[79] For Einstein, Maxwell's radical new vision of the "field as an ultimate entity" was more than just a refinement of scientific concepts.[80] The field offered an actual physical analogue of the arcane philosophical idea that there was a single unified foundation for all things. It provided the perfect material vessel for the pantheist's monist vision.

Einstein was determined to discover a detailed mathematical model of this divine substrate, and his relentless pursuit of this grand goal reveals him at his most fanatical. No sooner had general relativity been confirmed by the eclipse expedition of 1919 than he began to dream even bigger. Just a few weeks after he'd become the most famous scientist in the world, he wrote to his friend Max Born, "I have not yet eaten enough of the fruit of the Tree of Knowledge."[81] His spectacular success had only made him hungry for more. "The relativity theory reduced to one formula all laws which govern space, time, and gravitation," he explained. "The purpose of my work is to further this simplification."[82]

It's no secret that Einstein's end goal was a nondualistic understanding of Nature. "For years, it has been my greatest ambition to resolve the duality of natural laws into unity," he explained to a London newspaper in 1929. He was convinced that "the force which moves electrons in their ellipses about the nuclei of atoms is the same force which moves our earth in its annual course about the sun, and is the same force which brings to us the rays of light and heat which make life possible upon this planet."[83] In a unified field theory, all physical forces would be reduced "to one and the same mathematical form," and all apparently individual objects would be understood as more or less intense emanations of a single underlying substrate.[84] "The particle would be merely a domain containing an especially high density of field energy," Einstein explained. "The disturbing dualism would have been removed."[85]

Einstein published his first unified field theory in 1923.[86] It was a fledgling effort, and he soon realized the theory wouldn't fly. But it wasn't long before he brought his full powers to bear on the problem, and by 1925 he thought he'd found "the true solution."[87] Within a few weeks, however, he realized that his latest brainchild was "beautiful but dubious," and after a couple of months he concluded that it too was "no good."[88] This pattern repeated itself over and over again. Almost every year saw another new theory, and one after another, all were found to be fatally flawed. Einstein was well aware of the troubling trend. "Most of my intellectual offspring," he admitted, "end up very young in the graveyard of disappointed hopes."[89] But he couldn't quit. In quiet desperation, he employed an increasingly exotic menagerie of mathematical methods: five-dimensional spacetime, affine connections, and even a new formalism he invented out of thin air and called "distant parallelism." Ultimately, his work on the unified field theory spanned nearly four decades and spawned some two dozen scientific papers—all of them essentially worthless.[90]

Einstein's relentless quest for a pure field physics seemed completely crazy to his contemporaries. Other members of the old guard, like Niels Bohr, founded flourishing schools that were fostering the next generation of great physicists. Meanwhile, brilliant members of the new wave like Wolfgang Pauli and Werner Heisenberg were busy building the basics of quantum mechanics. But Einstein, the one they all idolized and held in awe, had deserted them to dedicate himself to this delusional quest. "Many of us regard this as a tragedy," wrote his friend Max Born, "for him, as he gropes his way in loneliness, and for us who miss our leader and standard-bearer."[91]

But from Einstein's perspective, things looked very different. Finding a single substrate that could unify the whole universe was not just about tidying up some loose ends in physics or finding a slightly more elegant way of expressing what we already knew about Nature's laws. For Einstein, "unifying physics" was "the highest and most sacred duty."[92] To give up would have been apostasy, a dereliction of religious duty. It would have meant abandoning his faith that, fundamentally, all things were One. It's only a slight exaggeration to say that his soul was at stake.

As death drew near, Einstein became acutely aware that he would fail in this grand physical-philosophical project. And since he'd founded no school in physics and had no real followers, he knew he needed to appoint a successor, a kind of spiritual son to carry on his quest. Einstein was on a first-name basis with all the most brilliant physicists of his time, and any one of them would have been honored to be anointed as his intellectual heir. But one person in particular was the most promising of all, a prodigious mind that might perhaps succeed where even Einstein himself had failed.

Wolfgang Ernst Pauli (1900–1958) was born into a prominent intellectual family. His mother was a well-known writer; his father, a renowned pioneer in the field of protein chemistry; his godfather, the famous physicist and philosopher Ernst Mach. But even amidst this rather distinguished company, Pauli proved early on to be precocious. One story has it that at the age of twelve he attended a lecture given by the eminent theoretical physicist Arnold Sommerfeld and pointed out an error in one of the equations on the blackboard. A startled Sommerfeld admitted that "there I have indeed made a mistake."[93]

As it turned out, Pauli's childhood acumen was no anomaly. He published his first scientific paper (on general relativity, no less) while he was still in high school.[94] He earned his PhD in theoretical physics at twenty-one. And just a few years later, he made the decisive contribution to quantum mechanics that would eventually earn him the Nobel Prize: the law now known as the Pauli exclusion principle. He was only twenty-five years old—the same age Einstein had been when he'd done his seminal work on the photon and special relativity.[95]

Pauli's many innovations are hard to fathom for non-physicists, but as impenetrable as they are to the uninitiated, they form part of the very foundation of modern physics.[96] And when the time came for the scientific community to recognize Pauli's crucial contributions, it was Einstein who nominated him for the Nobel. When Pauli won the prize in 1945, Einstein, now an old man, made a moving speech proclaiming that he was passing the torch to this most promising mind. Einstein claimed that he was "at the end of his wisdom" and that "the great theory that encompasses all natural forces" would have to be completed by someone like Pauli.[97] "I will never forget the speech about me, and for me, that he gave," Pauli reminisced a decade later, on the occasion of Einstein's death. "It was like the abdication of a king, installing me as a kind of elected son, as his successor."[98]

Einstein could hardly have chosen a worthier champion to carry on the quest. Pauli understood that "Einstein's life ended with the challenge to us of a synthesis," and he didn't intend to disappoint his old mentor.[99] Much like Einstein, he spent years searching for a unified theory of physics. But unlike Einstein, who labored alone, Pauli collaborated closely with his old friend Werner Heisenberg. Despite many years of discouraging results, by 1957 both men were ecstatic about the advances they'd made and certain that success would come soon. Pauli believed they were about to complete "a beautiful theory that will light up the world," and he wrote ebulliently to Heisenberg that "it's all bound to turn out magnificently."[100] Heisenberg was equally enthusiastic. "I have attempted an as-yet-unknown ascent to the fundamental peak of atomic theory with great efforts during the last five years," he wrote to his sister. "And now, with the peak directly ahead of me, the whole terrain of interrelationships in atomic theory is suddenly and clearly spread out before my eyes."[101]

Before they'd published their theory or even submitted it for proper peer review, Pauli began to present their ideas publicly. Many of the world's most famous physicists flocked to see what towering intellectual edifice these two titans of physics had built together.[102] But it quickly became apparent that the theory's foundations were unsound. Pauli had always been a scathing critic of other people's sloppy thinking—some called him "the conscience of physics," while others spoke of him as "the scourge of God"—and now, as he presented a premature, poorly formulated theory in public, no one had any compunction about returning the favor.[103] As he stammered through his lecture, his fellow physicists tore the fledgling theory to pieces.

Pauli was crushed. "As I talked more and more, I believed in it less and less," he's said to have mumbled on the way home.[104] He soon penned a brusque letter to Heisenberg telling him he was backing out of the project and wouldn't permit his name to be attached to its publication.[105] To Pauli's dismay, Heisenberg forged ahead alone: a press release proudly proclaimed that he'd "discovered the basic equation of the cosmos" with "his assistant, W. Pauli."[106] But in the end, it didn't amount to anything. The dream team that had made so many key contributions to quantum mechanics came no closer than Einstein to creating a comprehensive model of the cosmos.

9

A MAGNIFICENT SYNTHESIS

Modern scientific theory is tending toward a sort of transcendental synthesis in which the scientific mind will work in harmony with man's religious instincts.

—**Albert Einstein, "Science and God" (1930)**

EINSTEIN'S ENTIRE LIFE WAS A WORSHIP OF ONENESS. FOR HIM, REAL science meant the unification of all physical phenomena; real religion meant unity among all living beings; and real revelation meant a transcendental synthesis of science and spirituality that would yield a union of the individual and the Infinite.

He knew he wasn't alone in this yearning for unity. "In spite of the catastrophic desire for power and luxury which characterizes our time," he thought, "the eternal goals of the human spirit are not yet forgotten."[1] Nowhere was this more apparent to him than in the fact that his own abstract theories had evoked so much awe and admiration. "This extraordinary interest which the general public takes in science today... is one of the strongest signs of the metaphysical needs of our time," he suggested. "It shows that people have grown tired of materialism, in the popular sense of the term; it shows that they find life empty and that they are looking toward something beyond mere personal interests."[2]

Einstein knew that the public's fascination had little to do with physics and even less to do with him in particular. Consciously or not, people could sense that the scientist now embodied the archetype of the seeker after truth. "Behind the tireless efforts of the investigator there lurks a stronger, more mysterious drive," as Einstein explained it. "It is existence and reality that one wishes to comprehend."[3] In the past, it was the prophets who had transmitted truth from on high, but now, physicists and mathematicians had become the conduits of higher states of consciousness. Equations were the new revelations.

Einstein saw nothing sacrilegious in this state of affairs. He'd always been convinced that "the intuitive and constructive spiritual faculties must come into play wherever a body of scientific truth is concerned."[4] All true science relied on something that transcended the simple facts of experience, something subtler and deeper than the raw data. "By this," he explained, "I mean that our moral leanings and tastes, our sense of beauty and religious instincts, are all tributary forces in helping the reasoning faculty toward its highest achievements."[5] And although he acknowledged that our "system of thought" was still "in a state of evolution," the general trend was unmistakable.[6] "Modern scientific theory is tending toward a sort of transcendental synthesis," he insisted, "in which the scientific mind will work in harmony with man's religious instincts and sense of beauty."[7]

For all his fanatical devotion to finding a unified theory in physics, even Einstein acknowledged another obligation: we also had to *expand* our physical theory until it could encompass biology and even psychology. We could never claim our task was complete until we comprehended the enigma of consciousness, and Einstein's hope was that someday we might create "a science which takes in the whole of reality"—a theory that finally acknowledged that "body and soul are not two different things, but only two different ways of perceiving the same thing."[8]

Einstein's ambition wasn't to explain away the human spirit by reducing consciousness to quarks and electrons. On the contrary, his overarching aim, as always, was integration. But he knew this was still only a distant dream. It had been three centuries since Spinoza had conceived "the majestic concept," as Einstein called it, "that thinking (soul) and extension (naturalistically conceived world) are only different forms of appearance"—in other words, that mind and matter were only different "conceptual interpretations of the same 'substance.'"[9] By Einstein's time, many deep thinkers agreed that the dualism of mind and matter had to be discarded. But despite a dedication to metaphysical monism in principle, little if any progress had been made in practice. "Even our age," Einstein sensed, "is still very far from grasping the full implications of this concept."[10]

The implications are indeed awe-inspiring. Einstein thought that if "the most general laws" of physics "were fully known," then we would "be able to deduce from them...the theory of every process of nature, including that of life itself."[11] Not that we were anywhere near achieving such an extensive elaboration of natural law. "I mean *theoretically*," Einstein continued, "because in practice such a process of deduction is entirely beyond the capacity of human reasoning."[12]

Biology provided a perfect example of how Nature's overwhelming complexity made a mockery of even our most comprehensive models. "Our necessarily primitive thinking," Einstein felt, "must inevitably prove inadequate to something as complex as a living organism."[13] But although living beings were far too complicated to be understood according to rigid rules, our inability to model biological processes with mathematical precision didn't mean that life failed to follow physical laws. The behavior of living beings was "beyond the reach of exact prediction because of the variety of factors in operation," Einstein insisted, "not because of any lack of order in nature."[14]

Likewise, Einstein thought we could be confident that mind also emerged from matter, even if we could never claim to prove it. These "more complex processes," he admitted, "cannot be represented by the human mind with the subtle exactness and logical sequence which are indispensable for the theoretical physicist."[15] All the same, he was convinced that "physics and psychology are only different attempts to link our experiences together by way of systematic thought."[16] For Einstein, it was only our own mental limitations that kept us from comprehending the continuity between mind and matter, rather than any lack of cohesion in Nature itself. "The fact that in science we have to be content with an incomplete picture," he insisted, "is not due to the nature of the universe itself but rather to us."[17]

Einstein wasn't especially optimistic about our ability to overcome our limitations. "Human nature always has tried to form for itself a simple and synoptic image of the surrounding world," he said, but he was disappointed with "how small a part of nature" could "be comprehended and expressed" in any design of this kind. "All that is subtle and complex," he concluded, "has to be excluded."[18]

Many of his fellow physicists disagreed. They were determined to go well beyond the traditional boundaries of their discipline, and in fact they made major progress in expanding physics into the farther realms of chemistry and biology. Max Born and Robert Oppenheimer, for instance, showed how the quantum wave function could be applied beyond single atoms to understand larger molecules—including the complex organic compounds from which living beings are composed.[19] Similarly, Erwin Schrödinger made the first really rigorous attempt to theorize about the physical structure of the genetic material within the living cell.[20] Both Francis Crick and James Watson independently acknowledged his insights as an important inspiration for their own investigations into the structure of DNA.[21] Einstein, on the other hand, remained focused on physics and refrained from doing any formal work in other fields.

He never made much of an effort to explore the world within, either. For all his genuine interest in the spiritual systems of the world, Einstein was always a bit of a dabbler when it came to the depths of the human mind. Although he admired the "serenity and selflessness" of the Buddhist monks he'd met in Asia, there's no evidence that he ever engaged in any traditional spiritual practices, such as meditation.[22] Nor did he bother to experiment with more modern methods of exploring the mind. When he was invited to undergo psychoanalysis, he flatly declined. "I should like very much to remain in the darkness," he replied, "of not having been analyzed."[23] He thought it was "possible that [psycho]analysis may paralyze our mental and emotional processes," and his considered opinion was that "it may not always be helpful to delve into the subconscious."[24] In a letter to Sigmund Freud, he even teased the father of psychoanalysis for never getting a chance to analyze him. "You, who got into the skin of so many people, and indeed of humanity, have had no opportunity to slip into mine," he wrote.[25] For all his insistence on the importance of integrating the psyche into our overall science of Nature, Einstein had little interest in exploring what he called the "unknown factors which mold our inner self."[26]

But unbeknownst to Albert Einstein, or almost anyone else in the scientific community, Wolfgang Pauli had been pursuing a synthesis of mind and matter for decades. Alongside his intimidating productivity in the scientific sphere, Pauli was carrying on another quest entirely: a fervent search for a "holistic union of psyche and physis."[27] Behind the public persona of the impeccable man of science, the fiercest critic and most fearless champion of quantum mechanics, Pauli was diving deep into his unconscious mind, trying to find "a new ('neutral') psycho-physical standard language" that would let him describe the "invisible, potential form of reality" that had given birth to both mind and matter.[28]

For the most part, Pauli kept these peculiar passions out of the public eye. Old friends like Werner Heisenberg were of course aware that under the "outward display of criticism and skepticism lay concealed a deep philosophical interest, even in those dark areas of reality or the human soul which elude the grasp of reason."[29] But Pauli was always reluctant to share his personal explorations of the inner realm publicly. Then as now, spiritual interests were usually considered incongruous, even unbecoming, in a serious scientist, and he feared being seen as some kind of mystic—or worse still, a madman.[30]

Pauli in particular had much to lose. In the scientific realm, there was no question that he'd fulfilled his early promise. The boy genius had done Nobel-worthy work by the time he was twenty-five years old. By the age of thirty, he was at the peak of his powers, both fêted and feared as one of the most penetrating minds in physics. In the eyes of the world, he was a colossus: one of the three kings of quantum mechanics, a protégé of both Niels Bohr and Albert Einstein, well on his way to becoming the unquestioned "conscience of physics."[31]

But in parallel with his meteoric rise to the pinnacle of science, in private Pauli was plummeting toward the depths of despair. Beginning in 1927, a series of tragic events struck in quick succession: his mother committed suicide; his father married a young mistress who was only Wolfgang's age; and Pauli's own brief marriage to a cabaret dancer ended in divorce. Aggravating it all, his alcohol intake, always substantial, had reached alarming levels. The onetime wunderkind was spending his nights in bars and brothels, getting into brawls and being ejected from the better restaurants. Incredibly, Pauli's scientific work proceeded at a phenomenal pace amidst all this emotional upheaval and unhealthy living, but his personal life was falling apart.[32]

In 1931, he turned to Carl Jung for help. Some twenty-five years Pauli's senior, Jung (1875–1961) had by this time become a wizard-like figure, a modern-day magus. He spent several months each year secluded in a kind of spiritual castle he'd built in the village of Bollingen, Switzerland, an intentionally archaic retreat without running water or electricity.[33] Here he immersed himself in ancient Greek and Latin spiritual literature and explored his inner world through a technique he called active imagination, whereby he consciously entered into dreamlike states while still awake.[34]

It didn't require any unusual acumen for Jung to see that Pauli was no ordinary patient. Jung immediately understood that he was dealing with an "extraordinary personality" whose dreams were "chock full of archaic material."[35] Intent on letting Pauli's unconscious unfold according to its own intrinsic impetus, he refused to analyze the physicist himself. Instead, "in order to avoid all personal influence," he pawned Pauli off on one of his protégés, a freshly trained analyst with almost no experience at all.[36] With Pauli thus insulated from Jung's own overpowering personality, Jung was convinced that "conditions were really ideal for unprejudiced observation and recording."[37]

He was astonished by what Pauli's unconscious unleashed. In a matter of months, Pauli produced a series of dreams that, in Jung's estimate, "could hardly be surpassed in intelligence, clarity, and consistency."[38] Jung saw Pauli as a virtuoso of the inner world, just as much a genius in exploring his own psyche as he was in the scientific realm. In what Jung called an "observation of the unconscious with minute accuracy," Pauli recorded more than a thousand dreams and even spontaneously illustrated many of them.[39] After just a few months of guidance, Pauli no longer "require[d] any assistance" in interpreting and amplifying on his inner material.[40] "He even invented active imagination for himself," Jung noted, amazed.[41]

Jung was so impressed that he ended up using Pauli's dreams as the basis for a detailed study of the unconscious. With Pauli's permission, he published them alongside his own ample commentary on how they provided a perfect exemplar of the process he called *individuation*: the development of the individual toward wholeness. The account is remarkable not just for the light it sheds on Jung's psychological methods, but also as one of the most intimate and in-depth accounts ever given of the inner world of a genius.[42]

Soon, Jung and Pauli were meeting in person on a regular basis—not so much for formal psychoanalysis, but rather for informal conversations that covered every conceivable topic.[43] It was the beginning of a close personal friendship and professional collaboration that would endure for almost three decades. Pauli had started off as just another analysand, but he soon became Jung's star research subject, then a close companion, a faithful correspondent, and finally a trusted critic of Jung's own creative work.[44] Eventually, the two of them even co-authored a book together,[45] and Pauli became one of the first patrons of the nascent C. G. Jung Institute in Zurich.[46]

And in a curious coincidence that looks a lot like kismet, it was none other than Albert Einstein who catalyzed this historic collaboration.

Jung remembered meeting Einstein in the "very early days," when the latter was still working on general relativity. Both men would soon be world-famous, but back then neither was well known, and they were able to sit down to a series of quiet dinners between 1909 and 1912. "I pumped him about his relativity theory," Jung recalled, but he knew he was "not gifted in mathematics" and acknowledged that he "felt quite small" when confronted with the colossal creations of Einstein's mind.[47] All the same, Jung claimed that Einstein's "genius as a thinker...exerted a lasting influence on my own intellectual work." It was Einstein who'd "first started me off thinking about a possible relativity of time as well as space and their psychic conditionality," he explained. "More than thirty years later, this stimulus led to my relation with the physicist Professor W. Pauli."[48] And so it was that Carl Jung, arguably the greatest living explorer of the inner world, entered into an intensive relationship with Wolfgang Pauli, Einstein's appointed successor in the scientific realm. It's exactly the kind of astonishing connection that Carl Jung never would have believed was merely a coincidence.[49]

This unusual intellectual-spiritual collaboration endured on account of a shared obsession: both men had a passionate desire to discover some way of merging mind and matter. Pauli considered "physics and psychology as complementary types of examination," and he was determined to develop a deeper theory that would not only fuse all the forces of Nature, but also finally unite matter and mind in a system he called "psycho-physical monism."[50] There could hardly have been a more perfect companion than Carl Jung, who was likewise on a quest to find "a psychic analogue of physical energy."[51] Together they spent some two decades searching for what Pauli described as "a synthesis embracing the rational understanding as well as the mystic experience of one-ness."[52]

The question, of course, was how to achieve such a synthesis. As a psychiatrist, Jung naturally placed the emphasis on inner experience. But for him this entailed much more than just psychoanalysis or self-exploration. Jung believed we could make contact with a mysterious wisdom within,

Einstein with Wolfgang Pauli in 1926. Pauli was widely considered the most brilliant and original physicist of the early twentieth century, second in stature only to Einstein himself. When Pauli won the Nobel Prize in 1945, Einstein, now an old man, claimed that he was "at the end of his wisdom" and declared that it was up to Pauli to discover "the great theory" that would unify all our knowledge of Nature. "I will never forget the speech about me, and for me, that he gave," Pauli wrote. "It was like the abdication of a king, installing me as a kind of elected son, as his successor." But Pauli considered "physics and psychology" to be "complementary types of examination." Collaborating closely with Carl Jung, he explored his unconscious mind in great depth and spent nearly three decades trying to discover a "unified conceptual foundation for the scientific comprehension of the physical as well as the psychical"—a "standard language" that would let him describe the "invisible, potential form of reality" that had given birth to both mind and matter. Photo by the physicist Paul Ehrenfest. © CERN, used with permission, courtesy of AIP Emilio Segrè Visual Archives.

an inner intuition that was often personified in our dreams and fantasies as a teacher who tried to enlighten us. Thus embodied as "the enlightener, the master and teacher," this inner wisdom could express "its meaning in the opinion and voice of a wise magician."[53]

Jung himself was no stranger to such experiences. He dreamt of many different teachers over the decades, but the one he thought of as his own inner master was the figure he called Philemon. Philemon first appeared to him in an imaginal experience of 1913 as "a winged being sailing across the sky." As Jung explained it, "Psychologically, Philemon represented superior insight....To me he was what the Indians call a guru...and the fact was that he conveyed to me many an illuminating idea."[54]

Jung was well aware that the encounter with an inner teacher was an age-old experience, familiar to spiritual seekers for thousands of years. Many mystics had made use of these personifications of higher wisdom to gain insights into the structure of their own psyches, or even to derive new codes of conduct for their communities. This motif of the inner guru was in fact so common that Jung came to think of it as an archetypal pattern inherent in the human psyche, an inner experience ubiquitous across all cultures. "Modern man," Jung argued, "in experiencing this archetype, comes to know that most ancient form of thinking."[55]

Wolfgang Pauli reconceptualized this archaic experience in the most revolutionary way imaginable. What if, he asked, instead of merely offering insights into our personal thoughts and feelings, this inner wisdom could also shed light on what was impersonal and fundamental? What if the unconscious could teach us not only about our own mental processes, but also about the basic matrix from which both mind and matter emerged? "Nothing is further from the thoughts of modern man than the idea of penetrating the secrets of matter in this way," Pauli wrote to Jung, "for he would actually rather use these symbols to penetrate the secrets of the soul."[56] But using the mysterious machinery of the unconscious to understand the ultimate foundation of the world was precisely what Pauli had in mind.

In a sense, physicists had been doing this for centuries. The many magnificent breakthroughs since the Scientific Revolution had shown without doubt that remarkable insights into the workings of reality could manifest in the human mind—that "pure thought can grasp reality," as Einstein had so memorably put it. What's more, many scientists agreed that inspiration was a process essentially independent of their own egos: subjectively, it felt as if the hidden order of reality somehow revealed itself in a merely human mind.[57] But what Pauli realized was that scientists had typically taken a passive approach to pondering their problems, waiting patiently for gifts from the goddess of wisdom to well up from within.

Pauli decided to adopt a different tactic. He would dive deliberately into his inner world and actually *discuss* the structure of reality with the wise figures he found in the depths of his unconscious mind. To him, it was only a question of refining our *inner* vision, the same way we refined scientific instruments to better observe the *external* world. In a remarkable comparison, he suggested that when a "yogin" practiced "meditation as a conscious activity," this deliberate spiritual practice was "analogous to the manufacture and application of the spectrograph in physics."[58] In the same way that increasingly precise experimental instruments allowed physicists to reveal a more refined picture of material reality, Pauli thought, "a methodically guided imagination" could result in "a raising or multiplying of consciousness."[59] His hope was that the higher states of consciousness that resulted could lead to a profound insight into the common substrate of both matter and mind.

Pauli dedicated himself to this inner discipline with his usual determination. "What is decisive for me is that I *dream* about physics," he declared, whereas others only "*think* about physics."[60] He meant this quite literally. Many of his dreams featured mathematical symbols, scientific laboratories, and experimental instruments.[61] Often a teacher would appear to explain the significance of the symbolism. "In my dreams there usually appears some figure of authority," he explained, "on the relevant special field of physics."[62] Pauli thought of these dream teachers as representing "a 'superior personality part,' which is more constant than the ego [and] can outlive it."[63]

Over the years, many such incarnations of inner wisdom offered Pauli instruction. In one remarkable dream, it was none other than Einstein who appeared and explained to Pauli that quantum mechanics was only a one-dimensional rendering of reality. A deeper description of things had to include another dimension so far ignored in all our scientific theories: the depths of the human psyche and the richness of inner experience.[64] After Einstein passed away, Pauli lamented that "a fatherly friend who was so kind to me is no more."[65] But actually, Einstein wasn't completely gone: a few weeks after Einstein died, Pauli once again dreamt of being in a laboratory with him conducting mysterious physical experiments. In describing the dream to Carl Jung, he had no difficulty interpreting his late mentor's presence. "I regard Einstein," he explained, "as a manifestation of the 'master.'"[66] The wise father figure lived on in Pauli's unconscious, giving him guidance even from beyond the grave.

As his imaginal life continued to unfold, Pauli realized he was being initiated into the unconscious by these inner teachers. He was certain his "fantasies" were "neither meaningless nor purely arbitrary," but rather formed "a basically coherent series" that he sensed had "the character of initiation rites."[67] But where was it all leading? For all the important insights promised by the unconscious, Pauli was beginning to dread going any deeper. In a letter to Jung, he confessed to his "fear of a sort of ecstatic state in which the contents of the unconscious... might burst forth, contents which, because of their strangeness, would not be capable of being assimilated by the conscious and might thus have a shattering effect on it."[68]

Jung understood all too well. When he'd first confronted the unconscious, he too had feared for his sanity. It was as if "the shadowy personifications of the unconscious," which were "irrational and incomprehensible to the person concerned," had "burst into the *terra firma* of consciousness like a flood."[69] Jung had been terrified of being overwhelmed by the primordial power of what he'd unleashed, and he understood that the ego, in self-defense, tended to retreat into its rationalizations and deny the value of inner experience. This "resistance of the conscious mind to the unconscious and the depreciation of the latter," Jung realized, was a "conflict" that would "appear again and again" in the individual's psyche.[70] "Such a conflict cannot be solved by understanding, but only by experience," he concluded. "Every stage of the experience must be lived through."[71]

As Pauli was learning firsthand, this resulted in a rather schizophrenic existence. "I swing from one extreme to the other," he confessed to Jung.[72] "In the first half of my life I was a cold and cynical devil to other people and a fanatical atheist and intellectual 'enlightener.'"[73] And now he was in danger of the opposite: of "becoming detached from the world—a totally unintellectual hermit with outbursts of ecstasy and visions."[74] He struggled to find a balance between these "diametrically opposed psychic attitudes" and the two very different ways of being they seemed to demand.[75]

Pauli knew this was not a dilemma he was facing alone. "My personal problem is also a collective one," he realized.[76] "This abrupt swing into the opposite is a danger not just for me but for our whole civilization."[77] And he was convinced that his own yearning for "a unified concept of the entire cosmos" was also a critical issue for the contemporary world.[78] "More and more," he wrote to Jung, "I see the psycho-physical problem as the key to the overall spiritual situation of our age."[79]

The solution, Pauli thought, was "to present the old views in a new form, one that befits our current scientific knowledge."[80] And yet he understood that "a monistic union of matter and soul" would have to be more than just the same old mystical insights forced into the mold of modern science.[81] Not that he had any desire to devalue or dismiss traditional mysticism. He was well versed in the world's religious literature and respected what spiritual experience could offer. "Mysticism seeks the unity of all external things and the unity of the inner man with them," as Pauli saw it. "This it does by seeking to see through the multiplicity of things as illusory and unreal. Thus there comes about, stage by stage, man's unity with the Godhead—Tao in China, Samadhi in India or Nirvana in Buddhism. The last-named states are likely to be equivalent from the western point of view to the extinction of ego-consciousness."[82]

Pauli acknowledged that all this was of immense value. But what was missing from traditional mysticism was the critical attitude of science. "Thoroughgoing mysticism does not ask 'why?'" Pauli pointed out. "It asks 'How can man escape the evil, the suffering, of this terrible, menacing universe? How can it be recognized as appearance, how can the ultimate reality, the Brahman, the One, the Godhead...be seen?'" In contrast, Pauli thought it was "in keeping with the spirit of western science" to ask tough questions: "Why is the One mirrored in the Many? What is it that mirrors, and what is mirrored? Why has the One not remained alone? What originates the so-called illusion?"[83]

Pauli fully intended to find answers to these questions. It was our "destiny," he declared, "to keep bringing into connection with each other these two fundamental attitudes, on the one hand the rational-critical, which seeks to understand, and on the other the mystic-irrational, which looks for the redeeming experience of oneness."[84] He knew he wasn't the first to attempt such a synthesis. It was an adventure, he thought, that "originates with Pythagoras in the sixth century B.C. [and] is then carried on by his disciples and developed further by Plato."[85] And now "the ancient doctrine of the Pythagoreans that numbers are the origin of all things and as harmonies represent unity in multiplicity" had found a new home in his own mind, in what Pauli called "the extremely Neopythagorean mentality of my unconscious."[86]

For Pauli, the "future development" of this doctrine would mean "an *extension of physics*" that would have to "continue until psychology can be adopted by physics."[87] But by no means was this some program of ruthless scientific reductionism aiming to strip reality of all meaning. On the contrary, Pauli thought that a unified framework "would not detract in any way from the reality of psychology."[88] This new understanding of Nature would be neither a reduction of subjective experience to sub-atomic events nor a rejection of the physical realm, but rather a "holistic union of psyche and physis"—a "unified conceptual foundation for the scientific comprehension of the physical as well as the psychical."[89]

Jung admitted to feeling overwhelmed by the enormity of this endeavor. "We are hovering here," he wrote to Pauli, "on the borders of what is conceivable."[90] But in Pauli's opinion, a successful synthesis *was* achievable. He sensed "a certain symmetry between matter and spirit," and it was precisely this affinity that made it possible to understand all things as part of a pure unity.[91] "The human spirit will always be able to create ideas fitting somehow to the external objects," he believed, "because both the human spirit inside and the perceived object outside are subjugated to the same cosmic order."[92]

This mysterious symmetry was the same idea advocated by Einstein of "a pre-established harmony manifested in cosmic laws and related to our minds."[93] In fact, it was an idea as old as the ancient Greek philosophers, and Pauli made it clear that he was pursuing exactly the same program. "With Plato's philosophy in mind," he wrote, "I should therefore like to suggest that the process of understanding nature...should be interpreted as a correspondence, a coming into congruence of inner images pre-existent in the human psyche with external objects and their behaviour. The bridge between sense perceptions on one hand and concepts on the other," he explained, "rests, according to this conception, on a cosmic order... embracing psyche as well as physis, subject as well as object."[94]

Pauli was referring to Plato's famous theory of *anamnesis*: that all knowledge is really only recollection. "I have heard wise men and women talk about divine matters," says Socrates in one of Plato's dialogues. "They say that the human soul is immortal; at times it comes to an end, which they call dying; at times it is reborn, but it is never destroyed." In other words, as individuals we were impermanent, but our essence was eternal. And in ambling across the aeons, our souls had already "seen all things" and "learned everything." Long before we'd become human beings and found ourselves bound in finite forms, we'd belonged to the Boundless, and nothing could ever abolish our affinity with Infinity. And so Socrates felt that "finding knowledge within oneself" was not as strange as it seemed, because "the truth about reality" was "always in our soul."[95]

In all probability, Plato owes the striking doctrine of *anamnesis*, as he owes so much else, to Pythagoras.[96] According to classicist Alister Cameron, "It was this legendary prophet of Number, standing midway between magic and mystery on the one hand and philosophical speculation on the other, whom Plato in his statement of the theory of Anamnesis chose to acknowledge as a spiritual and intellectual forebear."[97] And so Pauli, in consciously carrying on the Platonic tradition that all knowledge of Nature manifesting in our minds implied a kind of "mystic correspondence between macrocosm and microcosm," was only giving modern expression to an ancient Pythagorean idea.[98]

And yet Pauli knew that his psychophysical monism was at best a protoscience, "a form of description" still "only in a prescientific phase."[99] He warned that we "must not fall into the trap of assuming" that we already possessed anything "comparable with a well-formulated doctrine of scientific truths."[100] As he himself emphasized, "My *search* is for a process of conjunction (unification of opposites), but I have only partially succeeded in this."[101]

In his wisdom, Pauli was circumspect about the prospects of success. "Whether and when this *coniunctio* [conjunction] will be realized I do not know," he wrote to Jung.[102] "It is probably a long journey, one we are only just setting out on"—a voyage that might very well "require just as great an effort on the part of many people in what may be a distant future as has the development of science and technology in the last 300 years."[103] And Pauli knew that the past didn't provide much cause for optimism. "Taking a warning from the failure throughout the history of thought of all premature endeavours to achieve a unity," he wrote, "I shall not venture to make predictions about the future."[104]

Pauli's prudence would prove to be prescient. A few years later, after suffering agonizing stomach pains, he was admitted to the hospital and diagnosed with advanced pancreatic cancer. Within days he would be dead.[105] But even as the end approached, he was still yearning for unity. One of the last things he said was, "The only person I would love to see now is C. G. Jung."[106] But it was not to be. Pauli never saw Jung again, and within a few years Carl Jung too would be dead. The dream that had mesmerized so many profound minds over the millennia, from Pythagoras to Pauli, disappeared into the depths and was once again forgotten. A magnificent synthesis of mind and matter would have to wait.

10

THE ETERNAL ENIGMA

*We have been endowed with just enough intelligence to
be able to see clearly just how utterly inadequate that
intelligence is when we are confronted with what exists.*

—**Albert Einstein, letter to the
queen of Belgium, Sept. 19, 1932**

A TRANSCENDENTAL SYNTHESIS OF SCIENCE AND SPIRITUALITY WAS
the highest truth human beings could hope to achieve, the very
quintessence of Einstein's cosmic religion. If he'd known about Jung and
Pauli's quest to conquer duality, he no doubt would have applauded the
effort. And yet if he'd lived long enough to see them fail, it's doubtful
that he would have despaired. Einstein always insisted that the ultimate
ideal—whether we thought of it as knowing God, transcending duality, or achieving absolute Truth—was a task never finished, a goal never
reached, a journey never ended.

Einstein arguably contributed more than anyone else in history to
our understanding of the cosmos, but he was always humble about how
far the human mind had come—and how far it could go. "We see a

universe marvelously arranged," he said, "obeying certain laws, but we understand the laws only dimly."[1] Ultimately, the endless iterations of the arch-force would always escape our theories and formulations. "All our knowledge is but the knowledge of schoolchildren," Einstein thought. "Possibly we shall know a little more than we do now. But the real nature of things, that we shall never know, never."[2] We could cling to all kinds of comforting myths and models—Christianity, quantum mechanics, even the cosmic religion—but "before God," Einstein concluded, "we are all equally wise—equally foolish."[3]

Einstein's dying dream was to banish duality forever with a single unifying framework, and many of his contemporaries were convinced that this ultimate theory was just around the corner. Quantum physics had made such speedy progress, and had been so successful in predicting experimental observations, that it seemed obvious to them that *this* would be the cornerstone of all future physics. The foundation had been found; now it was only a question of continuing to build upward.

Einstein famously disagreed. "If I have learned anything from a long life's ponderings," he wrote right before he died, "it is that we are much further from a deeper insight into the elementary processes than most of our contemporaries believe."[4] He once claimed that he'd "thought a hundred times as much about the quantum problems as I have about general relativity," and after decades of deep reflection, he readily acknowledged that it was a real advance.[5] "There is no doubt that quantum mechanics has seized hold of a good deal of truth," he admitted, "and that it will be a touchstone for any future theoretical basis." But all the same, he insisted that "I do not believe that quantum mechanics can serve as a *starting point* in the search for this basis."[6]

Einstein had several serious qualms about quantum mechanics, and his various grievances evolved over the years.[7] But the single most important issue was that the new physics insisted that we could no longer speak of *objective* reality, at least not inside the atom. Niels Bohr's original model of the atom pictured it as a tiny solar system: electrons were actual entities that whizzed around the nucleus like planets orbiting the sun. Quantum mechanics crushed this quaint little model without mercy. Within the atom were neither waves nor particles, in fact nothing solid or certain whatsoever, but rather clouds of probability that somehow "collapsed" and became concretely real only when observed.[8] As Niels Bohr put it, "When it comes to atoms, language can be used only as in poetry."[9]

All this was anathema to Einstein. For him, "the belief in an external world independent of the perceiving subject" was "the basis of all natural science," and he was appalled at how quantum mechanics made "no claim to describe physical reality itself, but only the *probabilities* of the occurrence of a physical reality."[10] Einstein was well aware that his staunch adherence to the old idea of objective reality was seen by his colleagues as nothing more than "a metaphysical prejudice, empty of content."[11] But for Einstein, faith in an actually existing physical world was the foundation of all physics, the source of all scientific progress since the time of Pythagoras. He couldn't imagine abandoning his conviction that Nature was governed by an objective order.

Convincing the new quantum theorists was another matter entirely, of course. In the spring of 1926, shortly after Heisenberg published his seminal paper on matrix mechanics, Einstein invited him to sit down for a deep discussion of its implications.[12] Einstein charged the new quantum mechanics with committing the double crime of being both too naïve *and* too cynical. On the one hand, it exaggerated the importance of what was observed with experimental apparatus, tacitly assuming that these empirical data could be taken at face value. On the other hand, it underestimated the uncanny power of rigorous theoretical thinking to explain the unobservable essence of reality—of the potential of "pure thought" to "grasp reality."[13]

To hammer his point home, Einstein emphasized the obviously paradoxical position Heisenberg had put himself in. "In a cloud chamber we can observe the path of the electrons," Einstein reminded him, and yet "at the same time, you claim that there are no electron paths inside the atom."[14]* *Outside* the atom, everyone agreed that an electron was an actual observable entity; but *within* the atom, quantum mechanics claimed that it couldn't really be said to exist at all! "I have no wish to appear as an advocate of a naïve form of realism," Einstein told Heisenberg. "I know that these are very difficult questions."[15] But at the same time, he insisted that "this is obvious nonsense."[16]

This "obvious nonsense" is precisely the paradoxical perspective still embraced in physics today. Probability waves, the "collapse" of the wave function, and the ultimate limits on our knowledge of atomic processes all remain mainstays of modern quantum mechanics. And yet as crazy as it sounds, it's the most accurate and successful scientific theory ever devised. "For most of us quantum mechanics *is* the theory of everything," as one contemporary physicist has remarked. "Quantum mechanics explains the periodic table of the elements, the nuclear reactions that power the sun, and the greenhouse effect that leads to global warming."[17] What more could anyone want?

A lot more, if you're Albert Einstein. Right up until the end of his life, he never accepted quantum mechanics as the final word in physics. "The quantum-mechanical description of physical reality," he insisted, in one of his most infamous and controversial papers, "is not complete."[18] In private, he was decidedly less diplomatic. "This theory," he once ranted in a letter, "reminds me a little of the system of delusions of an exceedingly intelligent paranoiac, concocted of incoherent elements of thought."[19]

Einstein's opposition to quantum mechanics is often portrayed as pigheadedness, the "stubborn criticism" of an out-of-touch old man.[20] It isn't true. Although it's often forgotten, Einstein was actually one of the *founders* of quantum physics. It was Max Planck, of course, who'd first proposed the quantized nature of radiation in his historic lecture of 1900.[21] But it was Einstein who brought quantum theory into contact with concrete physical reality in his landmark paper on the quantized nature of light in 1905. In fact, his Nobel Prize was awarded for this key contribution to quantum physics, not for relativity.[22]

---

* A cloud chamber is a particle detector used by physicists to observe subatomic particles such as electrons.

The discovery of the photon was only Einstein's first foray into the quantum world; he continued making major contributions to quantum physics for another two decades. He was the first to propose the dual wave-particle nature of light; the first to suggest that there could be spontaneous "quantum leaps" in an electron's energy state; and the first to realize that quantum statistics implied the existence of a kind of quantum "liquid," a whole new state of matter.[23] In fact, as one contemporary physicist has noted, "it was Einstein who had introduced almost all the revolutionary ideas underlying quantum theory, and who saw first what these ideas meant."[24]

Surprising as this might sound, it's an uncontroversial statement. In light of his later opposition to quantum mechanics, it's easy to forget Einstein's enormous early contributions. But his record of revolutionary work speaks for itself, and the founders of quantum mechanics spoke for him, too. Max Born, Louis de Broglie, and Erwin Schrödinger all acknowledged Einstein as a major inspiration for the innovations that won them their Nobel Prizes.[25] Not only that, it was Einstein who nominated them (as well as Heisenberg) for the Nobel, and he didn't rest until they received it. He kept repeating his recommendations, year after year, until all four had won the award.[26]

These are not the actions of a blind and bigoted critic. On the contrary, no less a legend than Niels Bohr argued that Einstein's ingenious and carefully crafted critiques had been essential in advancing the new theory. "Einstein's concern and criticism," Bohr claimed, "provided a most valuable incentive for us all to reexamine the various aspects of the situation."[27] Although many of Einstein's criticisms ended up being unfounded, he forced the founders of the new theory to question every definition, scrutinize every data point, refine every experimental apparatus. And for this service, Bohr thought his fellow physicists should join him in "acknowledging the indebtedness of our whole generation for the guidance [Einstein's] genius has given us."[28]

In any case, all criticisms aside, it's not as if Einstein cavalierly dismissed the quantum description of reality altogether. "There is no doubt that quantum mechanics has seized hold of a beautiful element of truth," he believed.[29] "Probably never before has a theory been evolved which has given a key to the interpretation and calculation of such a heterogeneous group of phenomena."[30] But even as he admitted the accuracy of its observations and admired the explanatory power of its mathematical models, he flat-out rejected its philosophical implications. "Deep down it is wrong," he wrote, "even if it is empirically and logically right."[31] Einstein knew his attitude earned him the scorn of most other physicists. "I am well aware that our younger colleagues interpret this as a consequence of senility," he wrote.[32] "But an inner voice tells me that it is not yet the real thing."[33]

It seems to many physicists today that the uncertainty and unpredictability Einstein so abhorred are inalienable aspects of physical existence. But the triumphant proclamations about "proving Einstein wrong" are exaggerated.[34] Certainly, some of the specific features that Einstein found suspect, such as quantum entanglement (which he called "spooky actions at a distance"), have stood the test of time.[35] But these specific concerns were only small components of a more all-encompassing critique. To Einstein, all our knowledge of reality was "the result of an extremely laborious process of adaptation: hypothetical, never completely final, always subject to question and doubt."[36] Although quantum mechanics represented "an important, in a certain sense even final, advance in physical knowledge," like every other model of reality, it could only be provisional.[37] "The foundation," Einstein thought, "will be deepened or replaced by a more comprehensive one."[38]

History has come to characterize Einstein as the closed-minded one, but to him it was the proponents of quantum mechanics who refused to see the obvious: further progress required something truly novel. "Nature demands from us a synthesis," as Einstein put it, "which thus far has exceeded the mental powers of physicists."[39] In essence, Einstein wasn't claiming that *he* was right, but rather that *everyone was wrong* (himself included). "No doubt," he mused, "the day will come when we will see whose instinctive attitude was the correct one."[40]

For Einstein, this "instinctive attitude" was mostly a matter of modesty, a humble acceptance of the humiliating inadequacy of all our mental models. "There are no eternal theories in science," he said, and he warned that "one should beware of committing oneself too dogmatically to the present theory in searching for a unified basis for the whole of physics."[41] But genius and humility rarely go together, and Einstein understood that admitting ignorance was agonizing for the great minds of his time. "It is rather rough to see that we are still in the stage of our swaddling clothes," he wrote to Schrödinger, "and it is not surprising that the fellows [that is, Bohr, Heisenberg, Pauli, et al.] struggle against admitting it (even to themselves)."[42] In other words, the human mind was still in diapers—and those who proclaimed probabilistic, paradoxical models as the absolute apex of understanding only proved how immature we actually were. Einstein was convinced that the crowning of quantum mechanics as king of science had been premature. Quantum probability was a pretender to the throne, still too young to carry the royal mantle of Reality—and certain to be usurped sooner or later.

But by no means did Einstein see relativity as the rightful heir. "The theory of relativity," he admitted, "is nothing but another step in the centuries-old evolution of our science," and further evolution was inevitable.[43] "Every important advance brings new questions," Einstein pointed out. "Every development reveals, in the long run, new and deeper difficulties."[44] And yet there was no shame in the shortcomings of our science, so long as they were acknowledged—and eventually surpassed. "No fairer destiny could be allotted to any physical theory," Einstein thought, than that it should lead to "a more comprehensive theory, in which it lives on as a limiting case."[45]

Despite almost a century of further development in physics, Einstein's dying dream—a deep, unified understanding of our universe—still eludes us. Far from being unified, physics remains divided in two. General relativity is still used to describe things on the cosmic scale, where big distances and heavy masses mean that the force of gravity predominates. Conversely, quantum mechanics still covers the realm of the very small, where the other fundamental forces of Nature hold sway. Every physicist accepts the accuracy of the two theories within their respective realms, but neither relativity nor quantum mechanics has been able to take a single successful step beyond its own borders.

Einstein undoubtedly would have found this enduring duality disturbing—and just as he predicted long ago, every physicist now acknowledges the need for a new framework that can link the very large with the very small. But the current contenders for a "theory of everything" don't inspire much confidence. Contemporary theories are increasingly fanciful mathematical models without any grounding at all in physical reality. String theory, the dominant paradigm today, is built on the wild assumption that there are actually eleven dimensions of spacetime, rather than the familiar four we're acquainted with. The other major model, multiverse theory, goes even further than assuming extra dimensions: it argues that there are an infinite number of other entire *universes* out there, rather than the only one we know exists for sure. If only Wolfgang Pauli were still around, he'd probably point out that the path we're pursuing is "not even wrong."[46]

But wrong or not (or maybe not even wrong), Einstein probably would have applauded these ardent attempts to achieve unity. He knew better than anyone that finding a unified theory was a "Herculean task," not for the faint of heart.[47] And after forty years of his own failures, he felt that false starts and fantastical ideas could be forgiven. Detours and dead ends were unavoidable on the human mind's meandering ascent from ignorance to illumination, and in a way even our errors assisted the upward advance. "Creating a new theory," Einstein claimed, "is not like destroying an old barn and erecting a skyscraper in its place. It is rather like climbing a mountain, gaining new and wider views, discovering unexpected connections between our starting point and its rich environment. But the point from which we started out still exists and can be seen, although it appears smaller and forms a tiny part of our broad view gained by the mastery of the obstacles on our adventurous way up."[48]

We'd climbed quite a ways already, and the view wasn't bad from our present vantage point. But why claim, like the quantum physicists, that we'd scaled the summit? As far as Einstein was concerned, we were still fumbling around in the foothills, and there was no telling just how high the trail of truth might lead. "Unexpected adventures," he suspected, "still await us."[49]

Einstein's ability to admit the incomprehensible, even embrace it, once again sets his worldview apart from almost all other spiritual systems. Religions always seek closure: final laws and final prophets, final judgments and final destinations (some with nicer climates than others). But for Einstein, closure couldn't be expected given our short lives and our limited minds. "Brief is this existence, like a brief visit in a strange house," he once wrote. "The path to be pursued is poorly lit by a flickering consciousness whose center is the limiting and separating 'I.'"[50] Religions filled the void with reassuring fables, but for Einstein the real essence of spirituality lay in accepting our ignorance. "My feeling is religious," he explained, "insofar as I am imbued with the consciousness of the insufficiency of the human mind."[51] The only worthy response to the eternal enigma was to welcome it into our lives. "Let us accept the world as a mystery," Einstein implored.[52]

He didn't always heed his own advice. Despite his best intentions, the meaning of our mysterious existence haunted him. As death drew near, he admitted that "the Sphinx," the ancient Greek emblem of the riddle of existence, "does not let me free for a moment."[53] Even after all the triumphs he'd achieved in science, he knew truth couldn't be tamed so easily. "You imagine that I look back on my life's work with calm satisfaction," he wrote to a friend a few years before he died. "But from nearby it looks quite different. There is not a single concept of which I am convinced that it will stand firm."[54] He concluded sadly that "the Sphinx stares at me in reproach and reminds me painfully of the Uncomprehended."[55] Even with the magic wand of mathematics in hand, finding unity in multiplicity wasn't going to be easy. "The mathematical God smiles," Einstein said. "He knows how far we are from real depth of understanding."[56]

Einstein was in perfect agreement here with probably the only other person who could compare with his genius: Isaac Newton. Newton, for all his apparent (and actual) greatness, admitted that "to myself I seem to have been only like a boy playing on the seashore, and diverting myself in now and then finding a smoother pebble or a prettier shell than ordinary, whilst the great ocean of truth lay all undiscovered before me."[57] Einstein echoed his idol's resignation even as he honored his achievements. "Newton's theory of gravitation was probably the human mind's greatest attempt to explain natural phenomena," Einstein wrote. "But even he was restricted by the knowledge of his time. And I, too, cannot jump over my own shadow."[58] Einstein even expressed his own modesty with a similar seashore metaphor. "When man acknowledges that he can know all, but not just yet; when he has the humility to think himself not so important and considers himself only a grain on the shore of infinite wisdom, then," Einstein thought, "he is religious."[59] As Einstein's friend Max Born once said of him: "He knew, as did Socrates, that we know nothing."[60]

In light of our awesome ignorance, there are really only two basic attitudes available to us: reverence for the marvelous mystery that is existence, or resignation at the manifest meaninglessness of it all. Einstein opted for the first, of course, but despair seems to have become the default mindset of our postmodern milieu. It's not difficult to see why: it doesn't take much effort to dismiss the quest for truth as a childish delusion that deserves to be discarded. And the cynic certainly has every right to ask a trenchant question: who are scientists to tell us to revere reality, when it was science that sucked the sacred out of the cosmos in the first place? If God was dead, as Nietzsche had said, then it was physicists like Einstein who'd killed Him. It was their doubts, their discoveries, and their disbelief that had purged the universe of poetry and deprived us of the divine. Science showed us a universe of utterly inhuman dimensions in which the Earth seemed to vanish into insignificance. What role was there for feelings of reverence in a reality ruled by relentless natural laws? What possible relevance could the little lamp of consciousness have in this cold and colossal cosmos? Wasn't all human striving simply vanity?

Einstein's son Eduard certainly thought so. Father and son carried on a fascinating correspondence over many years, and in one letter to his illustrious father, Eduard offered a perfect epitome of the modern mind's malaise. "One should always remember," he wrote, "that humans are only one of an infinite number of living creatures that the Earth has already produced, that they will only be here for a relatively short time, and that their entire history, their progress is completely insignificant."[61] In a rather depressing diatribe, he declared *Homo sapiens* to be a "dubious species" and decided that there was "actually a desperately small difference between a genius and an idiot."[62]

To Eduard, history showed us no heroes of the intellect worthy of admiration, no higher ideals worth emulating. Even the greatest creations of the human spirit were worthless. "A work of art," he argued, "has no value in itself whatsoever. The only thing that can be said in its favor," he thought, was that it "helps [humans] through their terrible boredom." Then things got personal. "Science is even worse off," he continued scornfully. "It is totally useless." In a none-too-subtle stab at his father's lifework, he even went on the offensive against the fundamental Pythagorean program. "It is a serious affliction of human society," he wrote, "that it showers honors upon...the person who is possessed by a drive to get a grip on the phenomena of the natural world in terms of numbers and patterns."[63]

Einstein wasn't convinced. "I cannot agree with what you say," he replied. "It is, of course, an irrefutable point of view when one rejects values altogether—consistent pessimism or nihilism. But if one wants to value society and, beyond that, what is alive, and rejoices in the fact that consciousness exists, it is impossible not to acknowledge the highest stage of consciousness as the highest ideal."[64] For Einstein, the highest stage of consciousness was the state of contemplation that offered us a little glimpse of Infinity and of the great laws governing our glorious cosmos—and for him it was this that gave life meaning. "Without the sense of collaborating with like-minded beings in the pursuit of the ever unattainable in art and scientific research, my life would have been empty," Einstein confessed.[65] "I have toiled my entire life with [such] problems," he told his son. "The love for these things has never diminished in me and will remain with me until my final breath."[66]

And for Einstein, these feelings *didn't* diminish. If anything, they only increased in intensity. As he approached the end of his life, almost everything offered intimations of the Infinite: the dazzling complexity of even the littlest living creature reminded him of the enigmatic force inherent in all things. "When I am busy calculating and see an insignificant insect that flies on to my paper," he wrote, "then I feel...that for all our scientific magnificence we are miserable drops in the ocean."[67] He found himself forced to admit that "behind each cause is still another cause," and that "the end or beginning of all causes has yet to be found."[68] It was obvious to him that there was an indefinable *something* still unexplained by even our most sophisticated theories. "One only has to think about life in order to feel to the depths how clumsily primitive the whole of our science is," he wrote. "For life is preconceived in the atom, just as life resides in the fertilized egg; and the mystery of the whole is already contained in the lowest stages. If one manages to penetrate to even greater depths, then mathematics fail us, so that it becomes impossible to reckon what is implicit in the basic equations."[69]

That which was implicit in the equations, the ultimate ground of the "greater depths," was of course what believers had always called God. And although Einstein preferred to give it names like "the grandeur of reason incarnate in existence," by no means did he believe that a change of terminology alone would allow him to escape the mind-boggling mystery of being.[70] On the contrary, when a bold interviewer once asked him point-blank if he believed in God, Einstein confessed that he was baffled by this greatest of all questions. "Your question is the most difficult in the world," he replied. "It is not a question I can answer simply with yes or no. I am not an Atheist. I do not know if I can define myself as a Pantheist. The problem involved is too vast for our limited minds."[71]

And yet, all doubts aside, Einstein was certain that there was "something eternal that lies beyond the reach of the hand of fate and of all human delusions."[72] Beyond our mathematical models, beyond all imaginable experience, beyond the theories of theologians and philosophers, was "the mysterious force that sways the constellations."[73] And in trying to fathom this force, we found ourselves in a rather paradoxical position. "We have been endowed with just enough intelligence," Einstein observed, "to be able to see clearly just how utterly inadequate that intelligence is when we are confronted with what exists."[74]

But as competent, creative beings born from the fertile womb of a wondrous world, it was beneath us to lament our limited understanding and our lack of ultimate certainty. "One must not take oneself too seriously," Einstein wrote to Eduard. "For one is a critter that barely became two-legged via the ape, a short-lived snippet of consciousness heavily burdened with atavistic instincts."[75] It would be "sheer folly," he thought, "to ponder interminably over the reason for one's own existence or the meaning of life in general."[76] To be sure, accepting the eternal enigma with humor and humility was no easy task. But we would know we'd arrived when we found ourselves in that serene state of mind "where the world ceases to be the scene of our personal hopes and wishes, where we face it as free beings, admiring, questioning, and observing."[77]

True: our lives were brief, our brains limited, our biases legion. But regardless, Einstein believed we should be grateful for our many gifts and proud of our powers, however paltry. "It is enough for me," he said, "to contemplate the mystery of conscious life perpetuating itself through all eternity, to reflect upon the marvelous structure of the universe which we can dimly perceive, and to try humbly to comprehend even an infinitesimal part of the intelligence manifested in nature."[78] Because for Einstein, our finitude was also our versatility. All of Nature's immeasurable forces found fulfillment in our unbearably fragile physical forms. It was not a punishment but a privilege to be perched midway between microcosm and macrocosm, between the fleeting moment and fathomless eternity. Small enough to stand in awe of our infinite cosmos, yet large enough to enjoy the little things; conscious enough to contemplate our own mortality, and yet long-lived enough to feel a tender appreciation for a flower's ephemeral existence—truly, we found ourselves inhabiting a magical middle ground.

In one of his most famous poems, William Blake invited us "to see a World in a Grain of Sand, and a Heaven in a Wild Flower."[79] And Einstein, for one, welcomed the offer. "For if we had completely and in a scientific sense learned the processes in the grain of sand," Einstein said, then we would have understood "the most general laws of the universe, from which the quintessence of all other events would have to be deducible."[80] We didn't need to be given God's grace or force our way through Heaven's gates to appreciate existence in all its grandeur. The Infinite could be found even in the infinitesimal.

The wisest minds of all times have always agreed that Infinity can be apprehended even through the narrow aperture of human experience. Spinoza once wrote that "the mind is eternal insofar as it conceives things under a form of eternity."[81] This was what he called the "way leading to freedom," the route whereby we could "feel and experience that we are eternal."[82] As Schopenhauer saw it, Spinoza's aim was no different from the great goal advocated ages ago by the sages of ancient India: to see oneself "as Brahman, as the original being," to whom "all arising and passing away are essentially foreign."[83] Seen with an awakened mind, the Infinite was immanent in every instant; each moment was a medium through which we could find eternity.

True to form, Einstein put a scientific spin on this ancient wisdom. Although he admitted that "I cannot conceive of anything after my physical death," his denial of immortality in the childish sense of a soul reclining in the clouds didn't imply that we had no permanence.[84] It was clear to Einstein that "the capacity to build living bodies and consciousness is connected with matter."[85] Moreover, he maintained that both mind and matter were merely different modes of the same everlasting "substance of the universe."[86] Einstein envisioned this eternal substance as "neither solely material nor entirely spiritual," but something that transcended them both. "Man, too, is more than flesh and blood," he believed, "otherwise, no religions would have been possible."[87] And so Einstein concluded that "the mind is immortal in the same sense as the body"—not in the sense of a soul enduring forever in a blissful afterlife, but of a spirit that had seen its essence was indestructible.[88] This is why Einstein could say, echoing Spinoza, that "the truly religious man has no fear of life and no fear of death."[89] And why he could insist, without any insolence or self-importance, that "I do not need any promise of eternity to be happy. My eternity is now . . . In this sense I am a mystic."[90]

Just a year before he died, Einstein had a final conversation at his home in Princeton with the poet William Hermanns. Hermanns was teaching at Harvard at the time, and he brought along a young freshman physics major named Patrick to meet the great man. Hermanns took Einstein aside to tell him that "this young man was disillusioned with his studies and with the world in general and that he was close to suicide."[91] Patrick kept quiet during most of the visit, but as the conversation came to a close, he finally found the courage to ask his burning existential questions. "How will I know when I'm on the right path in my thinking?" he asked Einstein forlornly. "Is there anything worth believing?"[92] As Einstein escorted them out, he offered his eager young disciple a serene synopsis of his spirituality, an elegant little epitaph to his extraordinary existence. "This could all be a dream," he admitted. "You may not be seeing it at all. But you have to assume something. Be proud of being the mean between macrocosm and microcosm. Stand still and marvel."[93]

Einstein at his home in Princeton in 1954, a year before he died, with the poet and pacifist William Hermanns. Hermanns was a survivor of the Battle of Verdun during the First World War. The slaughter he saw there converted him into a spiritual seeker and a passionate pacifist. Einstein and Hermanns had several conversations about science and spirituality over the years, which Hermanns carefully recorded in his book *Einstein and the Poet*. The book remains the most complete record of Einstein's musings on the cosmic religion. Photo by Bill Miller, used with permission of Kenneth Norton and the William Hermanns Trust.

EPILOGUE

SUMMONS TO ASCENSION

If one doesn't play a part in the creative whole, he is not worth being called human. He has betrayed his true purpose.

—**Albert Einstein, in conversation with William Hermanns**

EVEN THOUGH EINSTEIN ENCOURAGED US TO STAND STILL AND MAR-vel at the majesty of existence, wallowing in Oneness was never enough for him. He always had more in mind than a mere melding of the individual and the Infinite. To Einstein, identifying as part of Infinity didn't mean dissolving into anonymous irrelevance or basking forever in heavenly bliss. Anyone who encountered the divine incurred a debt, an obligation to bring something back from the higher realm in order to benefit those of us still stuck in the world below.

Long ago, Plato promoted an analogous idea. In his famous allegory of the cave, he compared our everyday existence to being chained up in a cavern deep underground. There are no real things here: only imitation objects carved from wood and stone, lit from behind by a blazing fire that casts shadows of these ersatz objects upon the cavern wall. The cave symbolizes our consensus reality, the seemingly solid world of the senses—which to the philosopher was only a deceptive dream.[1] Most of us were condemned to spend our lives confined to the cave, but Plato also believed philosophy could free us from our chains. Through a severe discipline of the spirit, we could make our way "up the rough, steep path" and "into the sunlight" of "the world above," where we'd be granted a vision of the Good, that which "governs everything in the visible world and is in some way the cause of all things."[2]

Plato's story of "the upward journey of the soul" is the most famous philosophical fable of all time.[3] But its ending is far less familiar. For Plato, enlightenment isn't the end goal. Those who "make the ascent and see the Good" must not be allowed to stay in the sunlight of this "pure realm," but must "go down again to the prisoners in the cave and share their labors."[4] Those who visited the world above had to return to the cave in order to help others ascend and see the same vision.

Einstein agreed. Everyone who'd ascended out of the darkness had a duty to go back down to the world and disseminate the divine light. Our "highest destiny," he held, was not merely to *recognize* our affinity with the cosmos, but to "co-create with its laws."[5] Of course, it was easier to complain about the apparent cruelty of our cosmos and agonize over the absurdity of it all, or to follow the many enlightened minds who'd left it all behind to immerse themselves in the divine. But neither option appealed to Einstein. For him, the goal was to keep one's "gaze" focused "on eternal things" even as we took "an active part in everything" that was going on in "the human and temporal sphere."[6] This was undeniably a more demanding path. But in Einstein's eyes, to evade this obligation was to ignore the sacred task to which we'd all been summoned. "If one doesn't play a part in the creative whole," he thought, "he is not worth being called human. He has betrayed his true purpose."[7]

Einstein ascended on unusual wings. Whereas most other minds over the millennia had made the ascent through ardent spiritual practice or strenuous philosophical speculation, Einstein was borne aloft on the twin winds of science and mathematics. It might seem like an odd method of achieving mystical union, but there's no doubt that Einstein saw science as a means of communion between microcosm and macrocosm. And he didn't consider this capacity for communing with the cosmos a gift that he alone had been given. All the "great seers" of science, he thought, had been endowed with an "admirable faculty of getting into touch with the unity of Nature."[8]

Einstein tried to clarify the mechanics of this mysterious communion by making an example of the most magnificent mind in the history of science: Isaac Newton. For Einstein, people like Newton represented a quantum leap, a qualitatively different kind of mind. "Their importance to humanity," he argued, "is akin to the significance of animal life to the [inanimate] matter of the universe." These higher beings had gone beyond the mind's ordinary boundaries, but even more important, they'd brought something *back*—and they'd gladly given these gifts from the heavens to all humankind. "They are the bearers," Einstein believed, "of a higher level of consciousness."[9]

In a genius as original as Isaac Newton, it hardly even made sense to speak of him as a human being at all. "Such a man," Einstein thought, "can be understood only by thinking of him as a scene on which the struggle for eternal truth took place."[10] Seen in this way, the study of Nature was not simply the striving of a small part to appreciate the whole. It was *the cosmos contemplating itself*, the universe creating a mirror image of its own magnificent order and manifesting it in the human mind.

Perhaps Einstein conveyed this idea most clearly in a conversation recounted by the poet William Hermanns:

> "What do you think of Spinoza?" [Einstein] asked, and before I could answer he went on. "For me he is the ideal example of the cosmic man."...
> When I ventured to say that Spinoza had criticized the Bible too harshly, Einstein glowered. He retorted that I should read him more carefully, that Spinoza rejected metaphysical speculations and explained miracles as natural events, that he was deeply religious but divorced from dogma.

Spinoza's unorthodox theism had shocked me. I asked, "If God reveals himself in nature, why not in man?" "Have you never been awed by the power of man's rational mind?" Einstein quizzed. "And man's intuition, man's inspiration?"[11]

What Einstein was driving at was that the divine revealed itself just as much in the insights and experiences of the human mind as in the creativity and complexity of Nature. As Einstein explained it, the scientist who made a great discovery was "astonished to notice how sublime order emerges from what appeared to be chaos"—and felt forced to conclude that this order "cannot be traced back to the workings of his own mind."[12] For Einstein, this was proof of a "pre-established harmony manifested in cosmic laws and related to our minds," evidence that the "marvelous order" of the cosmos revealed itself "both in nature *and* in the world of thought."[13]

Spinoza had gone even further. For him, the human mind didn't just mirror this marvelous order; it was actually *identical* to the infinite intelligence. "The human mind is that same power of thinking, not insofar as that power is infinite...but insofar as it is finite," he explained. "The human mind" was "in this way part of an infinite intellect."[14] Arthur Schopenhauer agreed: "At bottom," he believed, "it is *one* entity that perceives itself and is perceived by itself."[15] And Einstein echoed both his idols almost exactly. "The soul given to each of us," he said, "is moved by the same living spirit that moves the universe."[16]

For Einstein, then, genuine inspiration meant aligning ourselves with the inexhaustible source that was inherent in every atom and incarnate in every animate being. And as the human mind mirrored the marvelous order of reality more and more accurately, it gained ever greater control over the vast energies lying latent everywhere in existence. Those who thought of Einstein as an atheist were surprised to hear him speak of scientific realizations as analogous to divine revelations. But Einstein himself saw no contradiction. "I never denied that mind channels power," he once protested. "It can be used in any way a man chooses, and the stronger man's faith in his mind, the more he can achieve."[17]

Arguably, no other scientist ever achieved so much. The incredible creations of Einstein's imagination continue to have an immense impact on science and society. Many of his theoretical predictions were so far-reaching that it's taken nearly a century for them to be confirmed. But mind-boggling discoveries keep corroborating his fundamental ideas and extending them in new directions, garnering one Nobel Prize after another for those who've followed in his footsteps.[18] Daily life is now dominated by devices derived in large part from Einstein's insights—including modern marvels like GPS satellites, solar panels, and semiconductors.[19] The transcendental revelations of yesterday are the technologies we take for granted today.

In most cases, it took so long to turn Einstein's abstruse theories into actual technology that he didn't live long enough to see the tangible impact his ideas would have on the world. But one fateful innovation found an application all too quickly. When Einstein published his famous equation $E = mc^2$ in 1905, he proved that an immense amount of energy was locked up in even minuscule quantities of matter. Back then, he was simply a young man doing physics in his spare time, a scientific seeker searching for unity in Nature and endeavoring to demonstrate the equivalence of matter and energy. But in just a few decades, his seemingly esoteric speculations had been turned into shockingly effective weapons.

Einstein always insisted that "I do not consider myself the father of the release of atomic energy," but it's obvious that he felt accountable for the aftermath.[20] Not long after the nuclear bombs were dropped on Hiroshima and Nagasaki, Leo Szilard—the man who'd first realized a nuclear chain reaction was really possible, and the very person who'd convinced Einstein to sign the letter to Roosevelt urging research into atomic weapons—went to visit Einstein in Princeton. Szilard remembers Einstein being deeply distraught by the end result of their efforts. "You see now that the ancient Chinese were right," Einstein told him. "It is not possible to foresee the results of what you do. The only wise thing to do is to take no action—to take absolutely no action."[21] It's an obvious allusion to the old Taoist notion of non-action, *wu wei*, which Einstein had encountered in his study of the Tao Te Ching.[22]

But non-action was never Einstein's forte, and so he found himself facing an ancient dilemma no thoughtful human being can evade. Every individual with spiritual inclinations eventually reaches a crossroads where a choice must be made between the *vita contemplativa* and the *vita activa*—and Einstein was no exception. He too had to grapple with the age-old conundrum of whether to withdraw and live a life of quiet contemplation, or instead immerse himself in the world's woes and work to relieve them.

It's the same problem Plato had pointed to: whether to remain in the bliss of the higher realm or return to the cave with wisdom from the world above. Einstein understood that this was a genuine dilemma, because what we brought back from the higher realm was by no means guaranteed to benefit humanity. The divine light could be blinding for those whose eyes were accustomed to the cave's darkness; and like little children who'd stolen fire from the gods, we were bound to get burned from time to time.[23] It was undeniable that tapping into these transcendental powers created the potential for both triumph and tragedy.

But for Einstein, the answer was not to withdraw from the world, much less to stop trying to touch the divine. "We must not condemn man," he argued, even if "his inventiveness" was so often "exploited for false and destructive purposes."[24] As Einstein saw it, "the real problem" was "in the minds and hearts of men."[25] We had to "remember that the fate of mankind hinges entirely upon man's moral development."[26] The crucial thing was that creativity had to be accompanied by compassion; scientific innovation had to be equaled by ethical enlightenment. Seen in this light, every gift from above was both "a great duty and a great opportunity," a chance to show the Infinite—and each other—that we were worthy of wielding our wondrous powers.[27] "Be creative," Einstein beseeched us, "but make sure that what you create is not a curse for mankind."[28]

Many paths lead to Infinity. By highlighting how Einstein and a handful of other mathematical mystics over the millennia have made this ascent, by no means do I wish to imply that this is the only way upward.

214

Still less do I wish to suggest that science and technology are the only genuine gifts from the gods.

I'm no mathematician myself, and I have only an amateur's understanding of the field of physics. My own approach to the Infinite has been via the inner insights of meditation, the sublime experiences of solitude in the wilderness, and the wonder that wells up in any scientist trying to fathom the baffling blob of brain matter that somehow gives birth to the marvel of the human mind. No two lives are alike, no two minds identical. Whether through mystical experience or mathematical equations, some mix of these methods or some other means entirely, every individual must find their own way of accessing the Infinite.

"Man has infinite dimensions," Einstein once said.[29] Mathematics might have been *his* mystical path, but he admitted other avenues were just as effective. He saw music, in particular, as an exceptional way of expressing Infinity, and once admitted in an interview that "if I were not a physicist, I would probably be a musician."[30] On a superficial level, science and music could hardly be more different. But Einstein felt that "both are nourished by the same sort of longing"—a longing to glimpse, and give voice to, the underlying order of the cosmos.[31] "Bach and Mozart are to me the creators of great impersonal harmonies; they are not the expressionists of personal experiences," Einstein explained. "They do not belong to that class of artists whose main striving is to give outer artistic expression to the internal Ego." Just like the great scientists, he thought, the great composers communed with the cosmos. "The contemplative spirit," Einstein claimed, "is in both."[32] He even proclaimed that "Mozart's music is so pure and beautiful that I see it as a reflection of the inner beauty of the universe."[33]

Not that music was the only alternative to mathematics. Far from it. Most of Einstein's heroes approached Infinity from totally different directions. Rabindranath Tagore entered the eternal through poetry. Wolfgang Pauli discovered the divine in the depths of his dreams. Baruch Spinoza found the Infinite in the intricate machinations of his own intellect. And Mahatma Gandhi found God in perhaps the least likely of places: mass political action.

For Einstein, all these various undertakings were equally valid expressions of the Infinite. Science and mathematics granted us extraordinarily powerful (and awfully ambivalent) gifts. But there were many other blessings we could bring back from beyond; the Infinite source offered gifts of the most varied kinds. Whether their work manifested as the uplifting sounds of a great symphony, the profound reflections of a philosophical treatise, or even the courage to stand by one's convictions, Einstein had enormous esteem for all "the great artists, ethical pioneers, and thinkers" who had helped "raise human society to a higher level of experience, vision, ethical being, and understanding."[34]

All of us had to find our own way of honoring Infinity, and the surface differences didn't negate the deeper identity of all these diverse endeavors. For Einstein, it wasn't the *means* of ascension that mattered, much less the mode in which we chose to express our own idiosyncratic encounters with the Infinite. What really mattered were the meta-values that guided us on the voyage—and that gave us the strength for the return journey.

For Einstein, these were not faith and belief, but rather wonder and humility, honesty and integrity. It's a stern form of spirituality, a discipline that doesn't allow for laxness in any domain. A respect for the spiritual dimension of human existence is not a license for superstition or sloppy thinking, any more than a respect for the rules of physical reality allows us to claim that atoms and energy are somehow more real than our own subjective experiences.

If anything, Einstein's teaching holds both science and spirituality to a higher standard than ever. Science has to humbly accept how far it still has to go, and admit that its conception of the cosmos will be incomplete until it can embrace consciousness. Spirituality has to come down from the ethereal realms where it's most comfortable, and commit to effecting actual change here on Earth. And some third force, some as yet unknown mode of thinking or method of intuition, must aspire to finally achieve a harmonious integration of mind and matter. If we imagine this as a task for tiny human minds, it sounds utterly impossible, a feat beyond the strength or wisdom of even the greatest genius. But before we despair, we should remember that we're surrounded and suffused by the source of all things—and that the gifts of the Infinite are inexhaustible.

As this book comes to a close, I'm well aware that I've circled endlessly around the center without ever entering the innermost sanctum. The most essential questions—how exactly we can reach the Infinite and what we can expect to find in this higher realm—remain unanswered. This omission was not so much an intentional choice as an inevitable outcome. In approaching anything as arcane as the essence of pantheism, the nature of knowledge, the origins of inspiration, or the expansion of consciousness, the written word can take us only so far. There is something ineffable in all this, something unsayable that can only be experienced on an individual level.

Socrates spoke of it as "the place beyond heaven" and insisted that "none of our earthly poets has ever sung or ever will sing its praises enough!" And yet even as he claimed that we could never say enough about its splendor, he also knew that nothing of its essential nature could ever really be conveyed. "What is in this place is without color and without shape and without solidity," Socrates says, an "intangible truly existing essence" that is "visible only to the mind."[35] For the ancient Greeks, it could be approached only through the mysterious process known as *noesis*, a means of knowing essentially independent of, and unrelated to, the intellect as we usually understand it. We might lament the fact that this ultimate experience cannot be expressed more clearly or accessed more easily. But as Socrates himself was forced to admit, "this is the way it is."[36]

I can only echo the ancients and offer yet another reminder: nondualism is not a new kind of knowledge, but a nuanced experience of the numinous everywhere. I mean this as both an admonition and an exhortation: as a warning against being satisfied with secondhand knowledge and also as a summons to seek this experience yourself. Plato once confessed that "there is no writing of mine about these matters, nor will there ever be one. For this knowledge is not something that can be put into words like other sciences."[37] But that didn't mean that *nothing* could be communicated about the things of greatest consequence. On the contrary, Plato believed that "after long-continued intercourse between teacher and pupil, in joint pursuit of the subject," something remarkable could happen. "Suddenly," he says, "like light flashing forth when a fire is kindled, it is born in the soul and straightaway nourishes itself."[38]

In the same way, our extended encounter with Einstein's mind might awaken something in *us*. But although Einstein's example can give us guidance, we shouldn't think of him as a guru. It's not for Einstein (much less me) to determine anyone else's spiritual path. No one has the right to decide for another the best approach to the Infinite—and no one else can dictate how best to implement, down here in the actual world, what one learns in the divine realm. What you hold in your hands is only a handbook pointing to these higher things—an enchiridion of the Infinite that highlights the striking potential of the human spirit. It is no substitute for actual experience. All interpretations and all explanations are as nothing compared to the *epopteia* that every soul must seek for itself.

Almost every spiritual teacher in history has asked us to seek this experience within. Einstein implored us to "look to the Heavens, and learn from them."[39] But to truly see oneself as a part of Infinity is to understand that, in the end, there's no difference. All things have lessons to teach us, and everywhere we look we see only ourselves. Everything is illuminated by the light of Infinity.

And yet there's every imaginable excuse to ignore the Infinite and remain within the comfortable confines of ordinary consciousness. Spiritual teachers will tell us we're unworthy; scientists will tell us we're insane. And a vicious voice within might very well whisper in our ear that our arrogant little ego has no right to encounter the Infinite, that this is an attainment only for saints and sages.

Ignore them all. Forget the fairy tales about the perfect spiritual paragons of the past. Despite all the religious propaganda about full enlightenment and infallible prophets, probably no one has ever perfectly embodied Infinity. And it doesn't matter. Total transcendence is unnecessary, because even a transient experience of the eternal has enormous transformative potential. Any encounter with the Infinite, however fleeting, initiates an inner metamorphosis. All subsequent experience is endowed with subtle new shades of meaning; all subsequent action is held to a higher standard; and all subsequent thought must now be measured against the magnificent wisdom of the empyrean.

Infinity is immanent everywhere and accessible, in principle, to all. But no one can ever be fully purified, or fully prepared, for an encounter with the Infinite. Accept it. Embrace imperfection as the starting point in any ascent. The lotus flower ascends and emerges into the sunlight upon a stem that stretches down into the darkest depths.

ACKNOWLEDGMENTS

This book was conceived, researched, and then almost completely written while I was in the middle of medical school, and then finished during my first year of medical residency. This is not an experience I regret, but it's certainly not one I would repeat—and hardly one I can recommend. In managing the mania that made it all possible, I tried constantly to keep in mind the wise saying of Socrates in Plato's *Phaedrus*: "The best things we have come from madness."

Many people worried that such an undertaking would lead me to fail out of medical school, or maybe even drive me to madness. On the contrary, it was the only thing that kept me sane. Writing this book gave me an excuse to turn away, once in a while, from the endless suffering that saturates the surface of existence and instead immerse myself in the ancient and honorable search for the Infinite.

But this book, and the ideas that form its basis, were much more than just an escape. When sleep deprivation made me depressed, when godawful multi-day exams made me grumpy, when the never-ending display of death and disease made me despair, the exploration of Infinity reminded me, over and over, of my real reasons for practicing medicine: the kinship of all creatures, the sacredness of life, the unity of all existence.

It also offered an excuse to learn from, work with, and befriend some of the brightest, most thoughtful people I've ever had the good fortune to meet. First and foremost: my agent, Anna Ghosh, saw the potential of this project from the beginning and took a chance on an unknown, unpublished author who sent her an unsolicited email. Her enthusiastic support provided encouragement at exactly the moment it was most needed. She was the book's first and fiercest champion, and she shepherded it every step of the way—from wild idea to polished proposal to finished manuscript. When we eventually got offers from publishers, she even arranged things so that *I was offered my first book deal on my birthday*. That's the kind of agent she is. Thank you for everything, Anna. Here's to many more books together in the coming years.

Another person absolutely essential to this project was my editor at Basic Books, Thomas "TJ" Kelleher. TJ likewise saw the book's potential from the beginning, and from the very start he gave his complete commitment. He offered countless invaluable suggestions, saved me from many embarrassing blunders, and gave me permission to insert my own personal story into the book. Maybe most important of all, he kept me focused on the central message when things started to spiral out of control. Thank you for all your input, TJ; the book is infinitely better for it (pun intended).

Also at Basic Books, I want to thank Kristen Kim, assistant editor, for her comments on the book's first draft, as well as Lara Heimert, senior vice president, for believing in the book and taking a chance on a first-time author. I also want to thank the many other staff at Basic who helped with the copyediting, production, and publicity. Finally, I want to thank the art department for putting up with my never-ending agonizing over the book's cover design, and for indulging the arbitrary aesthetic preferences of the author.

As I began writing my first book, I was blessed to have many seasoned authors hold my hand. Alex Soojung-Kim Pang provided advice early on that was both wise and timely. I can't thank him enough for encouraging a first-time author to formulate and then actually finish a book. And I'll never forget his sage advice: "Done is better than perfect," he told me. I'll let the reader judge the truth of that little gem.

Philosopher Evan Thompson has been a longtime friend, collaborator, and mentor—and also an inspiration as an author of numerous books focused on the intersection of science and spirituality. But I'd mainly like to thank him for helping me snag his literary agent, Anna Ghosh, for myself.

Acknowledgments

Journalist and author extraordinaire Jo Marchant also offered early encouragement and was very generous with her time in answering nuts-and-bolts questions about the writer's craft, as well as providing me with a broader picture of the publishing industry.

Dr. David Fideler, expert on all things Pythagorean and pantheist, answered my questions, ranging from the esoteric to the idiotic, about ancient Pythagoreanism and was also kind enough to read the sections on ancient Pythagoreanism closely and check for errors. Of course, any errors that remain are my own responsibility.

My old friend and former roommate Louis Kosak was the very first person to read the very first draft and provided welcome encouragement and insightful comments at the very earliest stages. Nicholas Dingwall, formerly a teacher of mathematics, was my roommate during the first few months of writing this manuscript. I want to thank him for putting up with my madness when the muse struck. He tolerated the same Vivaldi concerto being blared over and over again at full volume, and accepted my disturbingly antisocial tendencies while I slaved away on this crazy side project instead of studying medicine. I'm also grateful to him for comments on the final draft of the book, as well as for our in-depth discussions about whether mathematics really represents the inherent structure of the universe or is merely a fabrication of the human mind. TBD!

Books have strange beginnings. This one began in a cabin in the jungle on the Big Island of Hawaii, and I want to thank Kathryn and Kenneth, my hosts at the beautiful rainforest retreat where this book was born. I brought Ken Wilber's anthology *Quantum Questions* with me to this cabin in the jungle, but the other key document that convinced me there was an incredible story here that simply had to be told was a poorly formatted Excel spreadsheet. And for this I must thank the staff at the Albert Einstein Archives at the Hebrew University of Jerusalem. First off, my thanks to Miriam Kutschinski for her timely assistance in providing me with a complete catalogue of the books in Einstein's personal library (the aforementioned spreadsheet). When, after more than two years of research and writing, I finally traveled in person to the Einstein Archives in Jerusalem, I received a warm welcome from all the staff. Chaya

Becker, Anna Rabin, Miriam Kutschinski (again), and Dr. Roni Grosz all provided invaluable assistance. They hunted down obscure quotes, dug up old letters, forced open rusty cabinets, translated Einstein's obscure marginalia from German to English, and found the handwritten original of Einstein's key essay on cosmic religion. Most important, they put up with my constant questions and incessant interruptions day after day. My deepest gratitude to the entire staff of the archives for their hospitality, and for the work they are doing in preserving all that remains of the magnificent mind that was Albert Einstein. I am also grateful to the archives as an abstract, impersonal entity, for allowing me to publish various unpublished materials I unearthed there, as well as allowing me to take, and publish, photographs of some of the remarkable books in Einstein's personal library.

Finally, this book is dedicated to my parents, who raised me to have an insatiable love of learning and an unslakable thirst for knowledge. I grew up in a home where overflowing bookshelves reached from floor to ceiling in almost every room. It was like living in a well-curated library that covered every topic imaginable, from science to spirituality, art to ancient civilizations. Even more than this intellectual influence early on, they always gave their blessing after I embarked on the unusual and circuitous journey I chose to take after I left home. Whether I was moving into a Buddhist monastery or matriculating at Stanford's School of Medicine, they've always been there offering their unconditional love and support. Thanks, Mom and Dad.

On a less personal level, a book of this kind also owes a great debt to previous scholars, who have spent decades preserving, cataloguing, organizing, and translating the enormous wealth of material related to Einstein. I especially want to thank Kenneth Norton, adopted son of William Hermanns and trustee of the William Hermanns Trust, who helped to publish Hermanns's handwritten notes of his conversations with Einstein over many years. The resulting book, *Einstein and the Poet*, is the single most important record of Einstein's unvarnished views on all things religious and spiritual, and I don't think *Part of Infinity* would have been possible without this singular resource. I'm also grateful for my personal discussions with Mr. Norton and for his ongoing efforts to preserve the writings and pacifist message of William Hermanns.

The author and editor Alice Calaprice also deserves special recognition for compiling a vast and varied collection of Albert Einstein's quotations, as well as painstakingly hunting down and documenting the original sources to confirm each quote's veracity.* She has done posterity a great service by identifying the countless letters, lectures, and lively discussions that are the sources for so many famous quotations. The striking quotations in her anthology provided some of the initial inspiration for writing this book. Her compendium will forever remain an invaluable resource for everyone interested in what Einstein himself actually said and thought.

Speaking of what Einstein actually said: countless crazy things are attributed to Einstein, usually without any source given and often without any justification whatsoever. This is not just a problem on the internet. Prior books on Einstein's spiritual side tend to be light on direct quotations and heavy on (wild) speculation and (mis)interpretation. My intention here was not only to present an almost unknown side of Albert Einstein, but also to let him speak for himself in his own often eloquent words. Hence extensive endnotes were a necessity.

Even more important than letting Einstein speak for himself is opening the door so that others can explore for themselves. I don't see this book as the final word on Einstein's spirituality, but only the first. By carefully documenting all my sources, my aim was not to drown the reader in my own erudition. Rather, my goal was to provide an important resource, and set a new benchmark, for serious scholars of Einstein's spirituality. There are countless topics I couldn't include here, endless avenues that weren't adequately explored. The spirituality of the great scientists has only begun to be understood. So many of their secrets are still buried in little-known letters and unpublished manuscripts. And as in Einstein's case, so much of what is well-known is still poorly understood. I'll echo Einstein and say that I have done my share. Further exploration is a job for scholars of the future.

---

* For the fascinating story of how she came to undertake such a thankless task, see Calaprice, "Alice Calaprice on Einstein."

The author at the Albert Einstein Archives in Jerusalem, browsing through a book of Rabindranath Tagore's mystical poetry. Photo by Chaya Becker and © the author (2023), used with permission of the Albert Einstein Archives.

BIBLIOGRAPHY

Africa, T. W. "Copernicus' Relation to Aristarchus and Pythagoras." *Isis* 52, no. 3 (1961): 403–409.

Albert, D. Z. *Quantum Mechanics and Experience*. Harvard University Press, 1992.

Aristotle. *The Complete Works of Aristotle*, vol. 2. Ed. J. Barnes. Princeton University Press, 1995.

Au-Young, S. N. *Lao Tze's Tao Teh King: The Bible of Taoism*. Huginn, Munnin, 1938.

Bair, D. *Jung: A Biography*. Little, Brown, 2003.

Ball, P. *Serving the Reich: The Struggle for the Soul of Physics Under Hitler*. University of Chicago Press, 2014.

Bohr, N. "Discussion with Einstein on Epistemological Problems in Atomic Physics." In *Albert Einstein: Philosopher-Scientist*, ed. P. A. Schilpp, 199–241. Tudor, 1951.

Born, M. *The Born-Einstein Letters*. 2nd ed. Macmillan, 2005.

Born, M. "Einstein's Statistical Theories." In *Albert Einstein: Philosopher-Scientist*, ed. P. A. Schilpp, 163–177. Tudor, 1951.

Born, M. "The Statistical Interpretation of Quantum Mechanics." Nobel lecture, 1954. www.nobelprize.org/uploads/2018/06/born-lecture.pdf.

Brian, D. *Einstein: A Life*. John Wiley & Sons, 1996.

Browne, L. *Blessed Spinoza: A Biography of the Philosopher*. Macmillan, 1932.

Browne, L. *The World's Great Scriptures*. Macmillan, 1946.

Bruno, G. *Cause, Principle, and Unity*. Trans. R. J. Blackwell. Cambridge University Press, 1998.

Bruno, G. *On the Infinite, the Universe, and the Worlds*. Trans. S. Gosnell. Huginn, Munnin, 2014.

Bucky, P. A. *The Private Albert Einstein*. Andrews and McMeel, 1992.

Burkert, W. *Lore and Science in Ancient Pythagoreanism*. Trans. E. L. Minar Jr. Harvard University Press, 1972.

Calaprice, A. "Alice Calaprice on Einstein: The Man Behind the Myth." Princeton University Press website, 2020. https://press.princeton.edu/ideas/alice-calaprice -on-einstein-the-man-behind-the-myth.

Cameron, A. *The Pythagorean Background of the Theory of Recollection*. George Banta, 1938.

Capra, F. *The Tao of Physics*. Shambhala Publications, 2010.

Bibliography

Capra, F. *Uncommon Wisdom*. Simon & Schuster, 1988.

Chapple, C. K. *Nonviolence to Animals, Earth, and Self in Asian Traditions*. SUNY Press, 1993.

Clark, R. W. *Einstein: The Life and Times*. World Publishing, 1971.

Crick, F. *What Mad Pursuit*. Basic Books, 1988.

Cybulska, E. "Freud's Burden of Debt to Nietzsche and Schopenhauer." *Indo-Pacific Journal of Phenomenology* 15, no. 2 (2015): 1–15.

Dawkins, R. *The God Delusion*. Houghton Mifflin, 2006.

Debernardi, A., et al. "Alcmaeon of Croton." *Neurosurgery* 66 (2010): 247–252.

De Broglie, L. "Recherches sur la théorie des quanta." Doctoral dissertation, University of Paris, 1924.

Desai, A., and G. Vahed. *The South African Gandhi: Stretcher-Bearer of Empire*. Stanford University Press, 2015.

Descartes, R. *Meditations on First Philosophy*. Hackett, 1993.

Deussen, P. *The Philosophy of the Upanishads*. Trans. A. S. Geden. Gyan Press, 2021.

Deutsch, E., and R. Dalvi. *The Essential Vedanta: A New Source Book of Advaita Vedanta*. World Wisdom, 2004.

Dillon, J., and J. Hershbell. *Iamblichus: On the Pythagorean Way of Life*. Scholars Press, 1991.

Diogenes Laertius. *Lives of the Eminent Philosophers*. Trans. P. Mensch. Oxford University Press, 2018.

Dukas, H., and B. Hoffmann (eds.). *Albert Einstein: The Human Side*. Princeton University Press, 1979.

Einstein, A. *The Collected Papers of Albert Einstein*, vol. 14. Ed. D. K. Buchwald et al. Princeton University Press, 2015.

Einstein, A. *The Collected Papers of Albert Einstein*, vol. 15. Ed. D. K. Buchwald et al. Princeton University Press, 2018.

Einstein, A. Contribution to *Baruch Spinoza: Addresses and Messages Delivered and Read at the College of the City of New York on the Occasion of the Tercentenary of Spinoza, November 3rd, 1932*. Spinoza Institute of America, 1933.

Einstein, A. *Einstein on Cosmic Religion and Other Opinions and Aphorisms*. Dover, 2009.

Einstein, A. *Einstein on Peace*. Ed. O. Nathan and H. Norden. Avenel Books, 1981.

Einstein, A. *Einstein on Politics: His Private Thoughts and Public Stands on Nationalism, Zionism, War, Peace, and the Bomb*. Ed. D. E. Rowe and R. Schulmann. Princeton University Press, 2007.

Einstein, A. "Emission and Absorption of Radiation in Quantum Theory." *Proceedings of the German Physical Society* 18 (1916): 318–323.

Einstein, A. *Ideas and Opinions*. Modern Library, 1994.

Einstein, A. *Letters on Wave Mechanics: Correspondence with H. A. Lorentz, Max Planck, and Erwin Schrödinger*. Ed. K. Przibram. Philosophical Library, 2015.

Einstein, A. *Letters to Solovine: 1906–1955*. Philosophical Library, 2010.

Einstein, A. "On the Development of Our Views Concerning the Nature and Constitution of Radiation." *Physikalische Zeitschrift* 10 (1909): 817–826.

Bibliography

Einstein, A. "On the 200th Anniversary of Isaac Newton's Death" (1927). In *The Collected Papers of Albert Einstein*, vol. 15, ed. D. K. Buchwald et al., doc. 506. Princeton University Press, 2015.

Einstein, A. *Out of My Later Years*. Philosophical Library, 1950.

Einstein, A. *The Philosophy of Albert Einstein: Writings on Art, Science, and Peace*. Ed. W. Martin and M. Ott. Fall River Press, 2013.

Einstein, A. *The Principle of Relativity*. Dover Publications, 1952.

Einstein, A. "Remarks to the Essays Appearing in This Collective Volume." In *Albert Einstein: Philosopher-Scientist*, ed. P. A. Schilpp, 663–688. Tudor, 1951.

Einstein, A. "Science and God: A Dialogue." *The Forum* 83, no. 6 (1930): 373–379. https://archive.org/details/sim_forum-and-century_1930-06_83_6.

Einstein, A. *The Travel Diaries of Albert Einstein: The Far East, Palestine, and Spain, 1922–1923*. Ed. Z. Rosenkranz. Princeton University Press, 2018.

Einstein, A. "Über einen die Erzeugung und Verwandlung des Lichtes betreffenden heuristischen Gesichtspunkt." *Annalen der Physik* 17, no. 6 (1905): 132–148.

Einstein, A. *The Ultimate Quotable Einstein*. Coll. and ed. A. Calaprice. Princeton University Press, 2011.

Einstein, A. "What I Believe." *Forum and Century* 84 (1930): 193–194. http://drjingma.com/blog/whatibelieve.

Einstein, A. *The World as I See It*. BN Publishing, 2007.

Einstein, A., and L. Infeld. *The Evolution of Physics: From Early Concepts to Relativity and Quanta*. Cambridge University Press, 2008.

Einstein, A., B. Podolsky, and N. Rosen. "Can Quantum-Mechanical Description of Physical Reality Be Considered Complete?" *Physical Review* 47 (1935): 777–780.

Elkana, Y. "Einstein and God." In *Einstein for the 21st Century: His Legacy in Science, Art, and Modern Culture*, ed. P. Galison, G. J. Holton, and S. S. Schweber, 35–47. Princeton University Press, 2008.

Enz, C. P. *No Time to Be Brief: A Scientific Biography of Wolfgang Pauli*. Oxford University Press, 2002.

Euclid. *The Thirteen Books of Euclid's Elements*, vol. 1. 2nd ed. Trans. with introduction and commentary by T. Heath. Dover Publications, 1956.

Ferguson, K. *The Fire in the Equations: Science, Religion, and the Search for God*. Eerdmans, 1995.

Ferguson, K. *The Music of Pythagoras: How an Ancient Brotherhood Cracked the Code of the Universe and Lit the Path from Antiquity to Outer Space*. Walker, 2008.

Feuerstein, G. *The Yoga Tradition: Its History, Literature, Philosophy and Practice*. Hohm Press, 2008.

Fields, R. *How the Swans Came to the Lake: A Narrative History of Buddhism in America*. Shambhala Publications, 1981.

Fitzgerald, T. *Discourse on Civility and Barbarity*. Oxford University Press, 2007.

Fölsing, A. *Albert Einstein: A Biography*. Trans. E. Osers. Viking, 1997.

Frank, P. *Einstein: His Life and Times*. Da Capo Press, 2002.

Freeman, C. *The Closing of the Western Mind: The Rise of Faith and the Fall of Reason*. Vintage Books, 2002.

Bibliography

Galilei, G. *Dialogue Concerning the Two Chief World Systems*. Trans. S. Drake. University of California Press, 1970.

Gandhi, M. *An Autobiography, or The Story of My Experiments with Truth: A Critical Edition*. Yale University Press, 2018.

Gandhi, M. *The Wit and Wisdom of Gandhi*. Ed. H. A. Jack. Dover Publications, 2005.

Garrow, D. J. *Bearing the Cross: Martin Luther King, Jr., and the Southern Christian Leadership Conference*. HarperCollins, 1986.

Gelfand, M. "Albert Einstein Library: From Princeton to Jerusalem." *American Journal of Information Science and Technology* 3, no. 4 (2019): 80–90.

Gies, F., and J. Gies. *Cathedral, Forge, and Waterwheel: Technology and Invention in the Middle Ages*. HarperCollins, 1994.

Goldman, R. N. *Einstein's God: Albert Einstein's Quest as a Scientist and as a Jew to Replace a Forsaken God*. Jason Aronson, 1997.

Gould, S. J. "Nonoverlapping Magisteria." *Natural History* 106 (1997): 16–22. https://web.archive.org/web/20190403152432/http://www.stephenjaygould.org/library/gould_noma.html.

Greenspan, N. T. *The End of the Certain World: The Life and Science of Max Born*. Basic Books, 2005.

Gutfreund, H., and J. Renn. *Einstein on Einstein: Autobiographical and Scientific Reflections*. Trans. P. Argyres. Princeton University Press, 2020.

Guthrie, K. S., and D. Fideler (eds.). *The Pythagorean Sourcebook and Library: An Anthology of Ancient Writings Which Relate to Pythagoras and Pythagorean Philosophy*. Phanes Press, 1987.

Guthrie, W. K. C. *A History of Greek Philosophy*, vol. 1. Cambridge University Press, 1962.

Hadot, P. *What Is Ancient Philosophy?* Belknap Press, 2004.

Halpern, P. *Einstein's Dice and Schrödinger's Cat*. Basic Books, 2015.

Hanfstingl, B., et al. "Assimilation and Accommodation." *European Psychologist* 27, no. 4 (2021): 320–337.

Heisenberg, W. *Across the Frontiers*. Harper & Row, 1974.

Heisenberg, W. *Physics and Beyond: Encounters and Conversations*. Harper & Row, 1971.

Heisenberg, W. *Physics and Philosophy*. Harper & Row, 1958.

Hermanns, W. *Einstein and the Poet: In Search of the Cosmic Man*. Branden Books, 1983.

Hijiya, J. A. "The *Gita* of J. Robert Oppenheimer." *Proceedings of the American Philosophical Society* 144, no. 2 (2000): 123–167.

Hitchens, C. *The Missionary Position: Mother Teresa in Theory and Practice*. Verso, 1995.

Hoffmann, B., and H. Dukas. *Albert Einstein: Creator and Rebel*. New American Library, 1972.

Holton, G. "Who Was Einstein? Why Is He Still So Alive?" In *Einstein for the 21st Century: His Legacy in Science, Art, and Modern Culture*, ed. P. Galison, G. J. Holton, and S. S. Schweber, 3–14. Princeton University Press, 2008.

Howard, D. "A Peek Behind the Veil of Maya: Einstein, Schopenhauer, and the Historical Background of the Conception of Space as a Ground for the Individuation of

Bibliography

Physical Systems." In *The Cosmos of Science: Essays of Exploration*, ed. J. Earman and J. D. Norton, 87–150. University of Pittsburgh Press, 1997.

Huffman, C. *Archytas of Tarentum*. Cambridge University Press, 2010.

Infeld, L. *Quest: An Autobiography*. Chelsea, 1980.

Isaacson, W. *Einstein: His Life and Universe*. Simon & Schuster, 2007.

Isaacson, W. *Einstein: The Life of a Genius*. Collins Design, 2009.

Isaacson, W. "The Light-Beam Rider." *New York Times*, Oct. 30, 2015.

Jammer, M. *Einstein and Religion: Physics and Theology*. Princeton University Press, 1999.

Jardé, A. *The Formation of the Greek People*. Routledge, 1926.

Jaspers, K. *Spinoza*. Harvest/HBJ, 1957.

Jones, R. H. *For the Glory of God: The Role of Christianity in the Rise and Development of Modern Science*, vol. 1, *The Dependency Thesis and Control Beliefs*. University Press of America, 2011.

Joost-Gaugier, C. L. *Measuring Heaven: Pythagoras and His Influence on Thought and Art in Antiquity and the Middle Ages*. Cornell University Press, 2006.

Jordens, J. T. F. *Gandhi's Religion: A Homespun Shawl*. St. Martin's Press, 1998.

Jung, C. G. *The Archetypes and the Collective Unconscious*. Trans. R. F. C. Hull. Princeton University Press, 1959.

Jung, C. G. *Jung on Active Imagination*. Ed. J. Chodorow. Princeton University Press, 1997.

Jung, C. G. *Memories, Dreams, Reflections*. Trans. R. Winston and C. Winston. Pantheon Books, 1973.

Jung, C. G. *Psychology and Alchemy*. Trans. R. F. C. Hull. Princeton University Press, 1993.

Kant, I. *Religion and Rational Theology*. Ed. and trans. A. W. Wood and G. di Giovanni. Cambridge University Press, 2001.

Kayser, R. *Albert Einstein: A Biographical Portrait*. Albert & Charles Boni, 1930.

Kayser, R. *Spinoza: Portrait of a Spiritual Hero*. Trans. A. Allen and M. Newmark. Philosophical Library, 1946.

Keltner, D., and J. Haidt. "Approaching Awe, a Moral, Spiritual, and Aesthetic Emotion." *Cognition and Emotion* 17, no. 2 (2003): 297–314.

Kiley, J. F. *Einstein and Aquinas: A Rapprochement*. Martinus Nijhoff, 1969.

Kingsley, P. *Ancient Philosophy, Mystery, and Magic: Empedocles and Pythagorean Tradition*. Clarendon Press, 1995.

Kingsley, P. *Catafalque: Carl Jung and the End of Humanity*. Catafalque Press, 2018.

Kingsley, P. "From Pythagoras to the Turba Philosophorum: Egypt and the Pythagorean Tradition." *Journal of the Warburg and Courtauld Institutes* 57 (1994): 1–13.

Kingsley, P. *A Story Waiting to Pierce You*. Golden Sufi Center, 2010.

Kragh, H. *Cosmology and Controversy: The Historical Development of Two Theories of the Universe*. Princeton University Press, 1996.

Kragh, H. "Max Planck: The Reluctant Revolutionary." *Physics World*, Dec. 1, 2000. https://physicsworld.com/a/max-planck-the-reluctant-revolutionary/.

Kuznetsov, B. G. *Albert Einstein*. Progress, 1965.

Lao Tzu. *Tao Te Ching*. Trans. R. Wilhelm and H. G. Ostwald. Arkana, 1985.

Bibliography

Lukács, O. "Carl Gustav Jung and Albert Einstein: An Ambivalent Relationship." *Journal of the History of the Behavioral Sciences* 56, no. 2 (2020): 115–132.

Lyons, A. S., and R. J. Petrucelli. *Medicine: An Illustrated History.* Harry N. Abrams, 1987.

Macalister, T. *Einstein's God: A Way of Being Spiritual Without the Supernatural.* Apocryphile Press, 2008.

Macdonell, A. A. *A History of Sanskrit Literature.* D. Appleton, 1900.

Maurois, A. *Illusions.* Columbia University Press, 1968.

McEvilley, T. *The Shape of Ancient Thought: Comparative Studies in Greek and Indian Philosophies.* Allworth Press, 2002.

McGrath, A. *A Theory of Everything (That Matters): A Brief Guide to Einstein, Relativity, and His Surprising Thoughts on God.* Tyndale Momentum, 2019.

McKeon, R. *The Philosophy of Spinoza: The Unity of His Thought.* Ox Bow Press, 1987.

McMahan, D. L. *The Making of Buddhist Modernism.* Oxford University Press, 2008.

Meier, C. A. *Ancient Incubation and Modern Psychotherapy.* Northwestern University Press, 1967.

Meier, C. A. (ed.). *Atom and Archetype: The Pauli/Jung Letters, 1932–1958.* Trans. D. Roscoe. Princeton University Press, 2001.

Merzbach, U. C., and C. B. Boyer. *A History of Mathematics.* 3rd ed. John Wiley & Sons, 2011.

Miller, A. I. *Deciphering the Cosmic Number: The Strange Friendship of Wolfgang Pauli and Carl Jung.* W. W. Norton, 2009.

Miller, A. I. *Imagery in Scientific Thought: Creating 20th Century Physics.* Birkhäuser Boston, 1984.

Mitchell, R. G. *Einstein and Christ: A New Approach to the Defense of the Christian Religion.* Scottish Academy Press, 1987.

Moore, W. *Schrödinger: Life and Thought.* Cambridge University Press, 1989.

Morrow, S. B. *The Dawning Moon of the Mind: Unlocking the Pyramid Texts.* Farrar, Straus and Giroux, 2015.

Moszkowski, A. *Conversations with Einstein.* Trans. H. L. Brose. Sidgwick & Jackson, 1970.

Nadler, S. *A Book Forged in Hell: Spinoza's Scandalous Treatise and the Birth of the Secular Age.* Princeton University Press, 2011.

Nadler, S. *Spinoza: A Life.* 2nd ed. Cambridge University Press, 2018.

Naydler, J. *Shamanic Wisdom in the Pyramid Texts: The Mystical Tradition of Ancient Egypt.* Inner Traditions, 2005.

Neugebauer, O. *The Exact Sciences in Antiquity.* Barnes & Noble Books, 1993.

Nixey, C. *The Darkening Age: The Christian Destruction of the Classical World.* Houghton Mifflin Harcourt, 2018.

Numbers, R. L. (ed.). *Galileo Goes to Jail and Other Myths About Science and Religion.* Harvard University Press, 2009.

Oppenheimer, J. R. "Oppenheimer on Einstein." *Bulletin of the Atomic Scientists* 35, no. 3 (1979): 36–38.

Pais, A. *Einstein Lived Here.* Oxford University Press, 1994.

Bibliography

Pais, A. *Subtle Is the Lord: The Science and the Life of Albert Einstein*. Oxford University Press, 1982.

Pauli, W. *Writings on Physics and Philosophy*. Ed. C. P. Enz and K. von Meyenn. Trans. R. Schlapp. Springer-Verlag, 1994.

Picton, J. A. *Pantheism: Its Story and Significance*. Archibald Constable, 1905.

Planck, M. *Where Is Science Going?* Trans. J. Murphy. Ox Bow Press, 1981.

Plato. *Plato: Complete Works*. Ed. J. M. Cooper. Hackett, 1997.

Popkin, R. H. "Spinoza, Neoplatonic Kabbalist?" In *Neoplatonism and Jewish Thought*, ed. L. Goodmann, 387–409. SUNY Press, 1992.

Purser, R. *McMindfulness*. Random House, 2019.

Radhakrishnan, S. (ed.). *The Principal Upanishads*. Oxford University Press, 1953.

Rowland, I. D. *Giordano Bruno: Philosopher/Heretic*. University of Chicago Press, 2008.

Runes, D. D. *Spinoza Dictionary*. Philosophical Library, 1951.

Russell, B. *A History of Western Philosophy*. Simon & Schuster, 1945.

Russo, L. *The Forgotten Revolution: How Science Was Born in 300 BC and Why It Had to Be Reborn*. Springer-Verlag, 2004.

Saucier, G., and K. Skrzypińska. "Spiritual but Not Religious? Evidence for Two Independent Dispositions." *Journal of Personality* 74, no. 5 (2006): 1257–1292.

Sayen, J. *Einstein in America: The Scientist's Conscience in the Age of Hitler and Hiroshima*. Crown, 1985.

Schilpp, P. A. (ed.). *Albert Einstein: Philosopher-Scientist*. Tudor, 1951.

Schopenhauer, A. *Parerga and Paralipomena*, vol. 1. Ed. and trans. S. Roehr and C. Janaway. Cambridge University Press, 2016.

Schopenhauer, A. *The World as Will and Representation*, vols. 1–2. Trans. E. F. J. Payne. Dover Publications, 1969.

Schrödinger, E. *My View of the World*. Trans. C. Hastings. Cambridge University Press, 1964.

Schrödinger, E. *Nature and the Greeks and Science and Humanism*. Cambridge University Press, 1996.

Schrödinger, E. *What Is Life? The Physical Aspect of the Living Cell*. Cambridge University Press, 1944.

Schweder, R. A. *Thinking Through Cultures: Expeditions in Cultural Psychology*. Harvard University Press, 1991.

Schweitzer, A. *Indian Thought and Its Development*. Beacon Press, 1960.

Schweitzer, A. *The Philosophy of Civilization*. Macmillan, 1950.

Seelig, C. *Albert Einstein: A Documentary Biography*. Staple Press, 1956.

Seidel, E. "Spinoza." *European Judaism: A Journal for the New Europe* 34, no. 1 (2001): 57–69.

Shankara. *Crest-Jewel of Discrimination (Vivekachudamani)*. Trans. S. Prabhavananda and C. Isherwood. Vedanta Press, 1975.

Shastri, G. *A Concise History of Classical Sanskrit Literature*. Oxford University Press, 1943.

Singh, S. *Big Bang: The Origin of the Universe*. HarperCollins, 2004.

Smith, H. W. *Man and His Gods*. Little, Brown, 1952.

Smith, W. D. *Hippocrates: Pseudepigraphic Writings*. E. J. Brill, 1990.

Bibliography

Smolin, L. *Einstein's Unfinished Revolution*. Penguin Press, 2019.

Solovine, M. *Démocrite: Doctrines philosophiques et réflexions morales*. Libraire Félix Alcan, 1928.

Spencer, C. *The Heretic's Feast: A History of Vegetarianism*. Fourth Estate, 1993.

Spinoza, B. *The Collected Works of Spinoza*, vol. 1. Ed. and trans. E. Curley. Princeton University Press, 1985.

Spinoza, B. *Correspondence of Benedict Spinoza*. Trans. R. H. M. Elwes. Wilder Publications, 2007.

Spinoza, B. *Ethics* (1674). In *Spinoza: Complete Works*, ed. and trans. S. Shirley. Hackett, 2002.

Spinoza, B. *Spinoza: Complete Works*. Ed. and trans. S. Shirley. Hackett, 2002.

Spinoza, B. *Treatise on the Emendation of the Intellect* (n.d.). In *Spinoza: Complete Works*, ed. and trans. S. Shirley. Hackett, 2002.

Stone, A. D. *Einstein and the Quantum: The Quest of the Valiant Swabian*. Princeton University Press, 2013.

Tagore, R. *The Religion of Man*. Martino, 2013.

Tähtinen, U. *Ahimsa: Non-Violence in Indian Tradition*. Rider, 1976.

Talmey, M. *The Relativity Theory Simplified and the Formative Period of Its Inventor*. Falcon Press, 1932.

Taylor, C. C. W. *The Atomists: Leucippus and Democritus*. University of Toronto Press, 2010.

Taylor, T. *The Six Books of Proclus on the Theology of Plato*. Kshetra Books, 2017.

Thompson, E. *Why I Am Not a Buddhist*. Yale University Press, 2020.

Thoren, V. *The Lord of Uraniborg: A Biography of Tycho Brahe*. Cambridge University Press, 1990.

Ustinova, Y. *Caves and the Ancient Greek Mind: Descending Underground in the Search for Ultimate Truth*. Oxford University Press, 2009.

Ustinova, Y. *Divine Mania: Alteration of Consciousness in Ancient Greece*. Routledge, 2018.

Uzdavinys, A. *Philosophy as a Rite of Rebirth: From Ancient Egypt to Neoplatonism*. Prometheus Trust, 2008.

Vallentin, A. *The Drama of Albert Einstein*. Doubleday, 1954.

Van Erkelens, H. "Wolfgang Pauli's Dialogue with the Spirit of Matter." *Psychological Perspectives* 24, no. 1 (1991): 34–53.

Viereck, G. S. *Glimpses of the Great*. Macaulay, 1930.

Waterfield, R. *Plato of Athens: A Life in Philosophy*. Oxford University Press, 2023.

Watson, J. D. *Avoid Boring People: Lessons from a Life in Science*. Knopf, 2007.

Weinberg, S. *The First Three Minutes: A Modern View of the Origin of the Universe*. Updated ed. Basic Books, 1993.

White, M. *The Pope and the Heretic: The True Story of Giordano Bruno, the Man Who Dared to Defy the Roman Inquisition*. HarperCollins, 2003.

Whitehead, A. N. *Process and Reality*. Corrected ed. Free Press, 1985.

Whyte, L. L. *The Unconscious Before Freud*. Anchor Books, 1962.

Wigner, E. "The Unreasonable Effectiveness of Mathematics in the Natural Sciences." *Communications in Pure and Applied Mathematics* 13, no. 1 (1960): 1–14.

Bibliography

Wilber, K. *Quantum Questions*. New Science Library, 1984.

Wixwat, M., and G. Saucier. "Being Spiritual but Not Religious." *Current Opinion in Psychology* 40 (2021): 121–125.

Woit, P. *Not Even Wrong: The Failure of String Theory and the Search for Unity in Physical Law*. Basic Books, 2006.

Yates, F. *Giordano Bruno and the Hermetic Tradition*. University of Chicago Press, 1964.

Zabriskie, B. "Jung and Pauli: A Meeting of Rare Minds." In *Atom and Archetype*, ed. C. A. Meier, trans. D. Roscoe, xxvii–l. Princeton University Press, 2001.

Zackheim, M. *Einstein's Daughter: The Search for Lieserl*. Riverhead Books, 1999.

Zhmud, L. *Pythagoras and the Early Pythagoreans*. Oxford University Press, 2012.

Ziolkowski, T. *The View from the Tower: Origins of an Antimodernist Image*. Princeton University Press, 2016.

NOTES

Introduction: The Sacred Science

1. Erwin Schrödinger, quoted in Wilber, *Quantum Questions*, 92; Max Planck, quoted in Wilber, *Quantum Questions*, 151.

2. Max Planck, quoted in Wilber, *Quantum Questions*, 152; Werner Heisenberg, quoted in Wilber, *Quantum Questions*, 56.

3. Much more will be said about all this throughout the book. But for anyone skeptical at the outset, a good place to start would be two excellent books by historian Kitty Ferguson: *The Fire in the Equations* and *The Music of Pythagoras*.

4. This is the classic folktale "The Treasure at Home," the antiquity of which is uncertain. The tale is found in one form or another throughout the world's folklore at least as far back as the stories of Rumi, from where it probably made its way into *The Thousand and One Nights* (see, e.g., "The Dream" in *Tales from the Thousand and One Nights* [Penguin Books, 1973], 328–329). But the same idea is also present in English folklore as "The Pedlar of Swaffham," and has been made the center of popular novels in recent times, such as Paulo Coelho's *The Alchemist* (Harper Perennial, 1988), which is where I first came across it.

5. "Sacred science" is my own (admittedly loose) translation. Although *hieros* translates unequivocally as "sacred" or "holy," the meaning of *logos* in ancient Greek is much more difficult to discern. The root word has come to mean "science" or "study of" in our modern vocabulary: hence bio*logy* (the science of life) and cosmo*logy* (the science of the cosmos). But an equally valid translation might be "sacred doctrine" or "sacred teaching." As far as I'm aware, the *hieros logos* is first referred to by the ancient historian Herodotus (fifth century B.C.E.), who thought of it as some kind of sacred legend Pythagoras had inherited from his Egyptian teachers (see *Histories* 2.81). The historian Hecataeus of Abdera (fourth century B.C.E.) similarly claimed that Pythagoras imported some kind of "sacred doctrine" from Egypt; see Burkert, *Lore and Science*, 219. Obviously, the ancient authors aren't sure what exactly this sacred science consisted of. But the modern Pythagorean scholar Cornelia de Vogel argues that "the ἱερός λόγος [*hieros logos*]" was not just "a complex of heterogeneous elements: philosophy, morals and rules of communal life." Rather, "Pythagoras' thought was one all-embracing philosophical conception" centered on "the doctrine of number and harmony." And since Pythagoras sought a unifying worldview that embraced both science and spirit, it was

only natural for him "to apply these principles, established as a divine order in the cosmos, also to man and society." See de Vogel, *Pythagoras and Early Pythagoreanism*, 12.

6. Probably the single best summary of Einstein's spirituality is a short essay entitled "Religion and Science," *New York Times Magazine*, Nov. 9, 1930. Another account of his views was given in a lecture ("Science and Religion") delivered a decade later at Princeton Theological Seminary—not exactly a low-key venue for the presentation of radical religious views. Both are in Einstein, *Ideas and Opinions*, 39–47.

7. A detailed discussion of how Einstein's spirituality was received in his own lifetime is in Chapter 5.

8. Einstein, *Ideas and Opinions*, 42.

9. As quoted in William Miller, "Death of a Genius," *Life*, May 2, 1955, www .sundheimgroup.com/wp-content/uploads/2018/05/Einstein-article-1955_05.pdf. A truncated version of the quote also appears in Einstein, *The Philosophy of Albert Einstein*, 53.

10. All quotes here are from Einstein, "On the 200th Anniversary of Isaac Newton's Death." The original German for the remarkable phrase "the bearers of a higher level of consciousness" is "die Träger einer höheren Stufe des Bewusstseins."

11. Albert Einstein to Eduard Einstein, between Nov. 14 and Dec. 12, 1926, in *The Collected Papers of Albert Einstein* (henceforth CP) 15, doc. 415.

12. From an open letter of solicitation requesting funds to help educate the public on the dangers of nuclear weapons, sent out by the Emergency Committee of Atomic Scientists, over Einstein's signature, in early 1947. In Nathan and Norden, *Einstein on Peace*, 403.

13. Einstein to Max von Laue, Mar. 19, 1955, in *Einstein on Peace*, 621.

14. From a remarkable open letter (well worth reading in its entirety) that Einstein wrote to Sigmund Freud, July 30, 1932, in *Einstein on Peace*, 190.

15. The tale of Pandora is told by the ancient Greek poet Hesiod (thought to be roughly contemporaneous with Homer) in his *Works and Days*, written down sometime around 700 B.C.E.

16. Einstein, "On the 200th Anniversary"; Einstein, *Ideas and Opinions*, 52.

17. Hermanns, *Einstein and the Poet*, 66.

18. Einstein, "The Real Problem Is in the Hearts of Men," *New York Times Magazine*, June 23, 1946; also in *Einstein on Politics*, 383.

19. Einstein, "Religion and Science," *New York Times Magazine*, Nov. 9, 1930; also in *Ideas and Opinions*, 41.

20. Hermanns, *Einstein and the Poet*, 55.

21. From a major public policy statement issued by the Emergency Committee of Atomic Scientists on Apr. 11, 1948. Einstein was an extremely active member of the committee and signed the statement. In Einstein, *Einstein on Peace*, 473.

22. From a conversation with Algernon Black in 1940, Albert Einstein Archives Document 54-834, 10. Also quoted in Goldman, *Einstein's God*, 117. The first part of the quote is also in Einstein, *The Ultimate Quotable Einstein*, 334.

23. From a eulogy Einstein wrote for his friend Walther Rathenau, foreign minister of Germany, after he was assassinated by right-wing extremists. In Pais, *Subtle Is the Lord*, 12.

1: An Unfinished Quest

1. Einstein, *The Ultimate Quotable Einstein*, 95.

2. See Einstein, "What I Believe."

3. Hermanns, *Einstein and the Poet*, 140; Einstein, "Physics and Reality" (1936), in *Out of My Later Years*, 64.

4. Isaacson, *Einstein*, 541. The text has come to be known as the Russell-Einstein Manifesto, released by Bertrand Russell on July 9, 1955, a few weeks after Einstein died. It sparked the famous Pugwash peace conferences, which continue to this day, and which garnered their organizer, Joseph Rotblat, a Nobel Peace Prize in 1995. The full text of the manifesto is at https://pugwash.org/1955/07/09/statement-manifesto/.

5. See Yair Rosenberg, "Einstein's Last Speech," *Tablet*, Apr. 16, 2013, www.tabletmag .com/sections/news/articles/einsteins-last-speech. He'd been working on expanding the speech to "urge the creation of a world government to preserve peace." See Isaacson, *Einstein*, 541.

6. Einstein, *The Ultimate Quotable Einstein*, 339–340.

7. Einstein, *The Ultimate Quotable Einstein*, 273.

8. Hermanns, *Einstein and the Poet*, 27.

9. Einstein, *Ideas and Opinions*, 41.

10. "Congratulations to Dr. Solf," in Einstein, *The World as I See It*, 27–28.

11. Descartes famously divided all of existence into two separate substances, matter (*res extensa*) and mind (*res cogitans*). To address the obvious problem that consciousness seemed awfully well integrated with physical bodies, and especially human brains, he proposed that the mind somehow mingles with the body through the pineal gland, which lies near the center of the brain. See Descartes, *Meditations on First Philosophy*.

12. Ignoring consciousness altogether was the dominant trend in science and psychology throughout much of the twentieth century, best embodied by the school of thought that called itself behaviorism. Behaviorism is dead and buried, but similar ideas can still be found today in theories like epiphenomenalism, as well as in the work of certain philosophers of mind, such as the late great Daniel Dennett (for instance, in his book *Consciousness Explained*, which critics lampooned with satirical titles like *Consciousness Ignored* and *Consciousness Explained Away*).

13. One of the most famous advocates of such a stance was the renowned Harvard evolutionary biologist Stephen Jay Gould. Gould developed a viewpoint he called "non-overlapping magisteria," in which he claimed that "souls represent a subject outside the magisterium of science." See Gould, "Nonoverlapping Magisteria."

14. Dukas and Hoffmann, *Albert Einstein: The Human Side*, 38.

15. Einstein, "Moral Decay" (1937), in *Out of My Later Years*, 9.

16. Einstein, *Ideas and Opinions*, 49.

17. Einstein, "Science and God," 375; Einstein, *Ideas and Opinions*, 52.

18. Einstein, *Ideas and Opinions*, 49.

19. Einstein, *Ideas and Opinions*, 49.

20. See, e.g., Saucier and Skrzypińska, "Spiritual but Not Religious?," and Wixwat and Saucier, "Being Spiritual but Not Religious."

21. Schweder, *Thinking Through Cultures*, 68.

22. Einstein, "My View of the World" (1934), in *Ideas and Opinions*, 18; Einstein, "On Freedom" (1940), in *Ideas and Opinions*, 35.

23. Einstein, "Autobiographical Notes," 5.

24. Weinberg, *The First Three Minutes*, 154. Weinberg won the 1979 Nobel Prize in Physics for his work on unifying electromagnetism with the weak nuclear force—in other words, for continuing Einstein's great quest to unify the forces of nature.

25. Einstein to Otto Juliusburger, Apr. 11, 1946, in Dukas and Hoffmann, *Albert Einstein: The Human Side*, 82.

26. Hermanns, *Einstein and the Poet*, 66, 68.

27. Much more will be said on Einstein's acquiescence to the ultimate mystery of existence in Chapter 10.

28. Quoted in Kragh, *Cosmology and Controversy*, 55.

29. Einstein, "Science and God," 378.

30. Hermanns, *Einstein and the Poet*, 66.

31. Einstein, *Ideas and Opinions*, 49.

32. Hermanns, *Einstein and the Poet*, 57.

33. Hermanns, *Einstein and the Poet*, 57.

34. Hermanns, *Einstein and the Poet*, 26.

35. Hermanns, *Einstein and the Poet*, 72–73.

36. Hermanns, *Einstein and the Poet*, 69.

37. For instance: "I have never admired any system that encourages a herd nature in man by suppressing his free will to choose for himself....As long as I can remember, I have resented mass indoctrination. I do not believe in the fear of life, in the fear of death, in blind faith." Quoted in Hermanns, *Einstein and the Poet*, 131–132. Einstein was speaking specifically about communism and Christianity, respectively, but he obviously intended the point to apply more generally.

38. Hermanns, *Einstein and the Poet*, 65.

39. Einstein, *The Ultimate Quotable Einstein*, 95. After he was cremated, even his ashes were scattered in secret so that *that* location couldn't become a place of pilgrimage, either.

40. Einstein, *The Ultimate Quotable Einstein*, 274.

41. Einstein, *Ideas and Opinions*, 13.

42. For instance, Hitchens, *The Missionary Position*, criticizes Mother Teresa as far more interested in promoting the Catholic Church than in actually helping the poor. Garrow's Pulitzer Prize–winning biography *Bearing the Cross* documents MLK's many extramarital affairs with both married and unmarried women. And in *South African Gandhi*, Desai and Vahed highlight Gandhi's racist attitude toward native Africans during his stint living in South Africa, and his support of continued racial segregation. For a quicker read on the last topic, see Soutik Biswas, "Was Mahatma Gandhi a Racist?," BBC News, Sept. 17, 2015, www.bbc.com/news/world-asia-india-34265882.

43. Although he was by no means a monster, Einstein could be incredibly callous when it came to the well-being of his first wife and his two sons. See, for instance, Isaacson, *Einstein*, 225–248. Another biographer points out that "his relationship with his sons was strained, a lifetime source of unhappiness for him," and that "both marriages, Einstein sadly admitted, were failures." Quotes from Goldman, *Einstein's God*, 125, 126. See also Alok Jha, "Letters Reveal Relative Truth of Einstein's Family Life: Documents Show 20th Century

Giant Was Generous, Affectionate—and Adulterous," *The Guardian*, July 11, 2006, www
.theguardian.com/science/2006/jul/11/internationalnews.

44. See Einstein, *The Travel Diaries of Albert Einstein*. See also Kristine Phil-
lips, "Albert Einstein Decried Racism in America. His Diaries Reveal a Xenophobic,
Misogynistic Side," *Washington Post*, June 13, 2018, www.washingtonpost.com/news
/retropolis/wp/2018/06/13/albert-einstein-decried-racism-in-america-his-diaries-reveal
-a-xenophobic-misogynistic-side/.

45. For instance, Einstein once wrote: "It strikes me as unfair, even in bad taste, to
select a few individuals for boundless admiration, attributing superhuman powers of mind
and character to them." In *The Ultimate Quotable Einstein*, 14. He once even chided himself
for this particular failing, in the midst of an effusive foreword he wrote for a modern edition
of Galileo's *Dialogue*. As he gushed over Galileo's genius, he realized that "in speaking this
way I notice that I, too, am falling in with the general weakness of those who, intoxicated
with devotion, exaggerate the stature of their heroes." Quote from Einstein's foreword in
Galilei, *Dialogue Concerning the Two Chief World Systems*, vii.

46. See Einstein, "What I Believe."

47. The alleged perfection of people like Jesus and Mohammed needs little comment, but worldly leaders, too, are often accorded the same status. Even to this day,
many consider the Pope to be officially "infallible." For a thought-provoking look at the
origins of papal infallibility—and its prospects in the twenty-first century—see Rebecca
Rist, "Should Catholics View the Pope as Infallible," The Conversation, Jan. 10, 2019,
https://theconversation.com/should-catholics-view-the-pope-as-infallible-109548.

48. Carl Jung, quoted in Lukács, *Carl Gustav Jung and Albert Einstein*, 121.

49. Einstein, *The Ultimate Quotable Einstein*, 528.

50. Oppenheimer in a 1935 letter to his brother Frank (also a physicist), quoted in
Hijiya, "The *Gita* of J. Robert Oppenheimer," 129.

51. From Plato's *Theaetetus* 174a, as translated by Benjamin Jowett, www
.gutenberg.org/files/1726/1726-h/1726-h.htm.

52. Like so many others, Einstein was forced to flee Germany when Hitler rose to
power. He never again saw his native land, and from then on lived the life of an exile.
The Nazis took the opportunity of his absence to seize his home and all his belongings
for the glory of the Third Reich.

53. Her name was Lieserl. Born out of wedlock, the baby was kept secret and given
up for adoption; she probably died soon after of scarlet fever. The secret was so well kept
that even Einstein scholars and biographers were unaware of Lieserl's existence until John
Stachel of the Einstein Papers Project discovered some letters between Einstein and his first
wife, Mileva, that revealed Lieserl's existence. See Isaacson, *Einstein*, 75–76, 86–88. For a
book-length treatment of the issue, see Zackheim, *Einstein's Daughter*.

54. See Isaacson, *Einstein*, 417–419.

55. Lina Einstein in Auschwitz and Bertha Dreyfus in Theresienstadt. See Pais,
Subtle Is the Lord, 12.

56. Tallying all the anti-Semitism Einstein experienced in his lifetime would be
both pointless and depressing, but as just one of the more egregious examples: Einstein received a staggering sixty-two nominations for the Nobel Prize before he finally
won it—a disgraceful delay due in large part to anti-Semitic prejudice. See Isaacson,

Einstein, 309–316, for the story. It took seventeen nominations in a single year (1922) for him to finally win it, and even then the prize was awarded not for his revolutionary work on relativity but rather for his discovery of the photoelectric effect. Given his countless contributions to physics and cosmology, Einstein probably deserved three or four prizes—maybe even more (see Andrew Delbanco, "Fantasy Physics: Should Einstein Have Won 7 Nobel Prizes?," *Huffington Post*, Dec. 6, 2017, www.huffpost.com/entry /einstein-fantasy-physics_b_4948045). The actual win, when it finally came, was such an anticlimax that Einstein didn't even bother to attend. Notified a few months ahead of time that he should plan to be in Stockholm for the occasion, he decided to skip the ceremony and instead went ahead with a world tour that took him all over Asia. For a full list of all the nominations he received over the years, see the Nobel Prize website: www.nobelprize.org/prizes/physics/1921/einstein/nominations/.

There were many threats on Einstein's life over the years, some more serious than others. A near assassination attempt took place in Berlin in 1930, when a disgruntled writer decided to wreak vengeance on Einstein for not returning a manuscript the aspiring author had sent him. He claimed he was going to set dynamite and demolish Einstein's entire apartment building. The story is humorously told in Hermanns, *Einstein and the Poet*, 19–22; Hermanns recalls that Einstein "seemed utterly lacking in the instinct for self-preservation and was scarcely aware that people could hate him. For Einstein, this threat was as unreal as the ether." As for the animosity of other scientists, it suffices to cite a single book, the deplorable diatribe *One Hundred Authors Against Einstein*, published in 1931. The volume posed as a serious challenge to relativity, but in reality it was a motley collection of ill-informed arguments and ad hominem attacks. As usual, Einstein took it all in stride. "If I were wrong," he retorted, then a hundred authors wouldn't have been necessary—"one would have been enough!" In *The Ultimate Quotable Einstein*, 170.

57. The exception was Antarctica. Aside from extensive travels all over Europe, Einstein set foot on the soil of Egypt and Palestine; Brazil, Uruguay, and Argentina; and Sri Lanka, Singapore, Malaysia, China, and Japan.

58. We will encounter each of these figures later in the book.

59. See Holton, "Who Was Einstein?," esp. 9. Kant's magnum opus was given to Einstein by a young Jewish medical student, Max Talmey, who occasionally dined with the Einstein family. Talmey also gave the young Einstein various other mathematical and philosophical works. According to Talmey: "Soon the flight of his mathematical genius was so high that I could no longer follow. Thereafter philosophy was often a subject of our conversations. I recommended to him the reading of Kant. At that time he was still a child, only thirteen years old, yet Kant's works, incomprehensible to ordinary mortals, seemed to be clear to him. Kant became Albert's favorite philosopher after he had read through his *Critique of Pure Reason* and the works of other philosophers.... For five years I had the good fortune of associating very often with the young mathematician and philosopher. In all these years I never saw him reading any light literature. Nor did I ever see him in the company of schoolmates or other boys of his age. He usually held himself aloof, absorbed in books on mathematics, physics, and philosophy." See Talmey, *The Relativity Theory Simplified*, 164–165. Einstein gave Kant considerable credit for helping him outgrow the rigid belief systems of mainstream religion: "Reading Kant,

I began to suspect everything I was taught. I no longer believed in the known God of the Bible, but rather in the mysterious God expressed in nature" (quoted in Hermanns, *Einstein and the Poet*, 9). Elsewhere Einstein ascribes his break with mainstream monotheism to the popular science books of Aaron Bernstein (also given to him by Talmey) that he read around the same time (Gutfreund and Renn, *Einstein on Einstein*, 157). But these two different tales are probably best seen as complementary sides of the same story rather than conflicting accounts. Talmey also quotes Einstein as saying that "Bernstein's work…has exerted a very great influence on my whole development" (163).

60. The note was written sometime in the early 1950s, shortly before Einstein died. Reproduced in Pais, *Einstein Lived Here*, 82.

61. See Gelfand, "Albert Einstein Library."

62. See Infeld, *Quest*, 312–313. Einstein himself confirmed this. "*Don Quixote* has long been one of my favorite books," he wrote in a letter to his sister, Maja, from Berlin on March 15, 1926 (*CP* 15, doc. 222a).

63. Einstein, "What I Believe."

64. Einstein to his (likely) lover Betty Neumann, Nov. 13, 1923, *CP* 14, doc. 151.

65. As just one example: Einstein was well aware of the risks posed by the kind of world government he advocated; he simply thought the benefits outweighed them. "Do I fear the tyranny of a world government? Of course I do. But I fear still more the coming of another war or wars. Any government is certain to be evil to some extent. But a world government is preferable to the far greater evil of wars." Einstein, *The Ultimate Quotable Einstein*, 272.

66. Einstein's cosmic religion was first revealed to the public in "Religion and Science," *New York Times Magazine*, Nov. 9, 1930; also in Einstein, *Ideas and Opinions*.

67. Einstein, *The Ultimate Quotable Einstein*, 360.

68. Quoted in Dawkins, *The God Delusion*, 15; Michael Weiss, "It Doesn't Take an Einstein: The Problem with Using Scientists' Words to Support Religious Beliefs," *Slate*, June 18, 2008, https://slate.com/human-interest/2008/06/the-problem-with-using-scientists-words-to-support-religious-beliefs.html.

69. Alberto A. Martínez, "Was Einstein Really Religious?," Not Even Past, Apr. 11, 2012, https://notevenpast.org/was-einstein-really-religious-0/.

70. Cardinal William Henry O'Connell, quoted in Isaacson, *Einstein*, 388.

71. For an excellent overview of the origin of these concepts and how they've been utilized in various ways, see Hanfstingl et al., "Assimilation and Accommodation."

72. Dawkins's thoughts on the letter originally appeared on the Richard Dawkins Foundation website, but the article no longer appears to be available; however, the text can still be found at https://web.archive.org/web/20140427025031/http://www.richarddawkins.net/news_articles/2012/8/15/albert-einstein-s-historic-1954-god-letter-handwritten-shortly-before-his-death. Emphasis added.

73. Isaacson points to Einstein's "belief in God, albeit a rather impersonal, deistic concept of God" (*Einstein*, 385). He makes similar claims in a lecture he gave on the topic at the Aspen Institute in 2007 (see notes to Chapter 5).

74. Louis Menand, "Reading into Albert Einstein's God Letter," *New Yorker*, Dec. 25, 2018, www.newyorker.com/news/daily-comment/reading-into-albert-einsteins-god-letter.

75. See Martínez, "Was Einstein Really Religious?" Emphasis added.

76. Gutfreund and Renn, *Einstein on Einstein*, 148.

77. Some such list of five is generally considered to form the fundamental branches of all philosophy. The number and nature of the fundamentals differ depending on whom you ask, but it's a helpful framework in which to approach any religious or philosophical worldview. The specific list I'm using comes from my own memory of the introductory lecture to my high school philosophy class, which obviously made an impression. Metaphysics makes it onto most lists, but I leave it out here because I think, in Einstein's case at least, that his few metaphysical beliefs are captured well in his epistemology and his soteriology, and don't need to be considered in a separate category. Speaking of soteriology, modern philosophy has generally tended to shy away from questions of salvation and leave such concerns to theologians. But ancient philosophy is deeply concerned with understanding, encouraging, and attaining some kind of transcendence. In fact, it's no exaggeration to say that transcendence, in one form or another, was not just the primary concern but really the raison d'être of all ancient philosophical systems. Those curious about how philosophy declined from a training regimen for transcendence to the pedantic discussion of mere ideas can begin by consulting two brilliant books: Hadot, *What Is Ancient Philosophy?*, and Uzdavinys, *Philosophy as a Rite of Rebirth*.

78. In a recent bestselling biography of Einstein, weighing in at over six hundred pages, only ten are devoted to his spirituality: see Isaacson, *Einstein*, 384–393. I don't mean to disparage the biography as a whole, however, which covers an incredible amount of ground in a breezy style and admirable detail. Similarly, a recent volume on Einstein's legacy for the twenty-first century allots a meager thirteen pages to "Einstein and God"; see Elkana, "Einstein and God." And in discussing Einstein's most significant essay on his spirituality, his 1930 *New York Times Magazine* article on cosmic religion, biographer Albrecht Fölsing notes with satisfaction that it went "deservedly unnoticed when it was republished in Germany." See Fölsing, *Albert Einstein*, 632.

79. Mitchell, *Einstein and Christ*; Kiley, *Einstein and Aquinas*.

80. McGrath, *A Theory of Everything*, 168, 182.

81. Jammer, *Einstein and Religion*. Of course, this isn't the *only* book on Einstein's spirituality. One slim little volume brings together a few of Einstein's original writings on cosmic religion, but by providing almost no context or commentary, it leaves the door wide open to the same old misunderstandings. See Einstein, *Einstein on Cosmic Religion*. There's also a recent book claiming to explain "the original spiritual but not religious guy," but it's little more than a collection of out-of-context quotations and rambling, uninformed commentary: see Macalister, *Einstein's God*. Macalister's book does have one cardinal virtue, however: it recognizes that Einstein's spirituality has much in common with Taoism, which, as we'll discover, is much more than mere coincidence. The books I found by far the most useful, and closest to the truth, are Goldman, *Einstein's God*, and Hermanns, *Einstein and the Poet*, but both are brief accounts that don't explore in any real depth.

82. Einstein's Pythagorean faith in a rational cosmos comprehensible with mathematics is never even mentioned. There's not so much as a hint of Einstein's deep conviction that knowledge could dissolve the delusion of our separateness, as

exemplified in the quote that serves as this book's title. And Spinoza's pantheism—the only ism Einstein ever expressed any real affinity for—is totally neglected. But Einstein praised pantheism repeatedly (see the detailed discussion in Chapter 5). He called Spinoza "one of the deepest and purest souls our Jewish people has produced" and even wrote a short poem praising him, which began: "How much I love that noble man! More than I can tell with words" (*The Ultimate Quotable Einstein*, 153, 467). Jammer alleges that "the only connecting link between Spinoza's philosophy and Einstein's…[is] determinism" (*Einstein and Religion*, 45), and this total misreading of the connection between Einstein and his favorite philosopher is an egregious error. As we'll see, what spoke most to Einstein wasn't Spinoza's determinism, but his insistence that God and Nature are one, or in other words, that the divine is everywhere. "In common parlance," as Einstein once put it, "this may be described as 'pantheistic'" (*Ideas and Opinions*, 286). Even more than this, Einstein claimed that Spinoza's treatment of mind and matter as ultimately emerging from the same substance was what he admired most about his philosophy; this will be discussed in detail in Chapter 8.

83. Jammer, *Einstein and Religion*, 8–9. To be fair, the ideas about the Koran are Jammer's paraphrase of another scholar, not his own argument, but he is clearly supportive of the assertion and discusses it approvingly (rather than critically, as one might hope and expect).

84. Einstein's experiment keeping kosher laws is detailed by his sister, Maja, and Einstein learned the small catechism at his Catholic primary school (Jammer, *Einstein and Religion*, 16, 20). But what relevance any of this has to the mature spirituality of the world's greatest scientist is never made clear. Einstein was appalled by Catholicism (as is clear from many quotations throughout this book), and he also attested that orthodox Judaism was repugnant to him. "The religion of the fathers, as I encountered it in Munich during religious instruction and in the synagogue, repelled rather than attracted me," he told Jewish historian Julius Katzenstein on Dec. 27, 1931, quoted in Isaacson, *Einstein*, 30.

85. Actually, even more than a hundred pages in my edition of Jammer's book. See Jammer, *Einstein and Religion*, 153–265. Einstein quote in Hermanns, *Einstein and the Poet*, 70.

86. See Jammer, *Einstein and Religion*, 236.

87. Einstein's enormous admiration for Schopenhauer will be discussed in due course (see esp. Chapter 2). The quote from Schopenhauer is mentioned in Howard, "A Peek Behind the Veil of Maya," 99. He cites Schopenhauer's *Manuscript Remains*, 1:14, as the source. The full quote doesn't appear in the printed version of Howard's article; I found it in the manuscript draft, available online at www.academia.edu /download/32488488/A_Peek_Behind_the_Veil_of_Maya-Manuscript.pdf.

88. The final third of Jammer's book is wasted on bizarre intellectual acrobatics, with Jammer bending over backward in a futile effort to reconcile Einstein's thought with every conceivable contention of Christian theology. In the introduction, he admits that "it is possible that [Einstein] would have rejected all of the arguments in chapter 3 if he were alive" (11)—a commendable concession, albeit an obvious understatement.

89. Einstein, *Ideas and Opinions*, 245.

2: The Third Phase

1. Hermanns, *Einstein and the Poet*, 68.
2. Einstein, *Ideas and Opinions*, 39–40.
3. Einstein, *Ideas and Opinions*, 39–40.
4. Einstein, *Ideas and Opinions*, 39–40.
5. Einstein to Maurice Solovine, Jan. 1, 1951, in Einstein, *Letters to Solovine*, 105.
6. Hermanns, *Einstein and the Poet*, 67.
7. Hermanns, *Einstein and the Poet*, 67.
8. Einstein, *Ideas and Opinions*, 40.
9. "The religions of all civilized peoples...are primarily moral [i.e., second phase] religions." Einstein, *Ideas and Opinions*, 40.
10. Einstein, *Ideas and Opinions*, 40.
11. Einstein, *Ideas and Opinions*, 50.
12. Einstein, *The Ultimate Quotable Einstein*, 223.
13. Hermanns, *Einstein and the Poet*, 68.
14. Hermanns, *Einstein and the Poet*, 83.
15. For those interested in the ferocious blowback (from both professional theologians and the rank-and-file faithful) to Einstein's publication of his views on cosmic religion and his denial of a personal God, see the selection of angry letters and public condemnations in Jammer, *Einstein and Religion*, 97–114. This will be discussed in detail in Chapter 5.
16. Hermanns, *Einstein and the Poet*, 67.
17. Hermanns, *Einstein and the Poet*, 106.
18. Hermanns, *Einstein and the Poet*, 65.
19. Hermanns, *Einstein and the Poet*, 106.
20. Einstein, *Ideas and Opinions*, 41.
21. Einstein, "Religion and Science," *New York Times Magazine*, Nov. 9, 1930; also in *Ideas and Opinions*, 39–43.
22. Hermanns, *Einstein and the Poet*, 27.
23. Einstein, *Ideas and Opinions*, 43.
24. Einstein, *The Ultimate Quotable Einstein*, 11.
25. Einstein, *Ideas and Opinions*, 41.
26. Einstein, *Ideas and Opinions*, 41.
27. Hermanns, *Einstein and the Poet*, 68.
28. Hermanns, *Einstein and the Poet*, 68.
29. Einstein, *Ideas and Opinions*, 42.
30. Einstein, *Ideas and Opinions*, 42–43.
31. Einstein, *Ideas and Opinions*, 52.
32. Einstein, *Ideas and Opinions*, 53.
33. Einstein, *The Ultimate Quotable Einstein*, 336.
34. Although not entirely unprecedented: as we'll see, Einstein did have an important precursor in the puzzling figure of Pythagoras (see Chapter 4). Einstein's aim needs to be distinguished from, for instance, the countless Christian scientists through the ages who've seen their exploration of the natural world as their homage to God's

Creation. There are similarities to Einstein's view, of course, but there is also a crucial difference. More on this at the end of Chapter 5.

35. Einstein, *Ideas and Opinions*, 41.

36. Hermanns, *Einstein and the Poet*, 68.

37. Much more will be said about Bruno and Gandhi in Chapters 5 and 6.

38. Einstein, *Ideas and Opinions*, 40.

39. Einstein's 1954 letter to Joseph Lewis, who had sent him a copy of his book *An Atheist Manifesto*. In *Einstein on Peace*, 611.

40. Einstein's foreword to Smith, *Man and His Gods*, ix.

41. Much more will be said about Einstein's reverence for the independent mind in the next chapter.

42. Arthur A. Macdonell explains that "after Alexander's invasion, the Greeks became to some extent acquainted with the learning of the Indians; the Arabs, in the Middle Ages, introduced the knowledge of Indian science to the West; a few European missionaries, from the sixteenth century onwards, were not only aware of the existence of, but also acquired some familiarity with, the ancient language of India; and Abraham Roger even translated the Sanskrit poet Bartrihari into Dutch as early as 1651. Nevertheless, till about a hundred and twenty years ago [i.e., 1780] there was no authentic information in Europe about the existence of Sanskrit literature, but only vague surmise, finding expression in stories about the wisdom of the Indians." Macdonell, *A History of Sanskrit Literature*, 1.

43. Schopenhauer had an amusing (and insightful and prescient) take on this situation: "We, on the contrary, now send to the Brahmans [i.e., Indians] English clergymen and evangelical linen-weavers, in order out of sympathy to put them right, and to point out to them that they are created out of nothing, and that they ought to be grateful and pleased about it. But it is just the same as if we fired a bullet at a cliff. In India our religions will never at any time take root; the ancient wisdom of the human race will not be supplanted by the events in Galilee. On the contrary, Indian wisdom flows back to Europe, and will produce a fundamental change in our knowledge and thought." See Schopenhauer, *The World as Will and Representation*, 1:357.

44. The lion's share of the credit for initiating this interest is usually given to William Jones—justifiably, in my opinion, or so at least it seems in the eyes of an amateur who can't claim a comprehensive appreciation of such a vast and complex field. For instance: "It was in the seventeenth century that the European people, particularly missionaries and travellers, came to know of the Indian languages.... It was, however, Sir William Jones who did most to arouse the interest of Europeans in Indian literature" (Shastri, *A Concise History of Classical Sanskrit Literature*, 1).

45. After Baruch Spinoza, undoubtedly Einstein's favorite philosopher.

46. On the first page of the first chapter, he wrote: "This basic truth [i.e., that the apparent multiplicity of the world is an illusion of the mind masking from us an underlying unity] was recognized by the sages of India, since it appears as the fundamental tenet of the Vedanta philosophy." Schopenhauer, *World as Will and Representation*, 1:3–4.

47. Schopenhauer, *World as Will and Representation*, 1:xv.

48. See Moore, *Schrödinger*, 112.

49. Einstein, *Ideas and Opinions*, 41. "In his little study [in his home in Berlin], he works beneath the pictures of three thinkers of whom he is especially fond, Faraday, Maxwell and Schopenhauer" (Kayser, *Albert Einstein*, 194).

50. "Above all, he read a lot of Schopenhauer...he often sat with one of the already well-worn Schopenhauer volumes, and as he sat there reading, he seemed so pleased, as if he were engaged with a serene and cheerful work...He always insisted that the engagement with Schopenhauer...gave him far and away greater pleasure than, for example, Goethe." Konrad Wachsmann (the architect who designed Einstein's summer home in Caputh), quoted in Howard, "A Peek Behind the Veil of Maya," 92.

51. Einstein is supposed to have said, "The religion of the future will be a cosmic religion...Buddhism answers this description." *The Ultimate Quotable Einstein*, 482. The quotation is unsourced and placed by Calaprice under the heading "Probably Not by Einstein." But even if Einstein didn't say or write these exact words, we know he felt something more or less along these lines from his praise (already quoted) of Buddhism as containing a "much stronger element" of the cosmic religious feeling.

52. Schopenhauer, *The World as Will and Representation*, 1:xv.

53. Not that all these effects are necessarily positive, of course. But it's undeniable that Eastern philosophy, broadly speaking—and Eastern meditation specifically—have saturated Western culture since they first became big in the 1950s. For two different takes on how all this happened (both brilliant), see Fields, *How the Swans Came to the Lake*, and McMahan, *The Making of Buddhist Modernism*. For some of the concerns and controversies, see Purser, *McMindfulness*.

54. See Capra, *Uncommon Wisdom*, 43.

55. Pauli, *Writings on Physics and Philosophy*, 138; Pauli to Carl Jung, Feb. 27, 1952, in Meier, *Atom and Archetype*, 74.

56. Quoted in Capra, *The Tao of Physics*, 18.

57. Oppenheimer studied Sanskrit while working as a professor of physics; see Hijiya, "The *Gita* of J. Robert Oppenheimer," 130. "By 1933 he was reading it [the *Gita*] in the original Sanskrit" (151), and he even translated the *Gita* into English himself (148). He also read many other classic Hindu texts, "albeit in English translation" (129).

58. "Erwin read everything written by Schopenhauer...many westerners first learned of Vedanta and the Upanishads through his writings. His direct influence on Schrödinger was considerable, but equally important was the introduction he provided to Indian philosophy." Moore, *Schrödinger*, 111–112. Elsewhere Moore, in describing Schrödinger's notebooks on Vedanta, writes that "his thought has taken wings into the realm of eastern mysticism. He was obviously strongly influenced by Schopenhauer, of whom he had read every word" (*Schrödinger*, 172).

59. Schrödinger, *My View of the World*, 21. This awakening or insight is disguised, in good cautious scientific style, as a kind of hypothetical thought experiment, but it bears all the signs of a personal spiritual experience. I came to the conclusion that this was a thinly veiled personal experience on my own, but I was later delighted to find that Schrödinger's biographer had arrived at the same judgment about this exact passage: "He does not claim that he himself has achieved this fusion of self with Nature, but the lyrical intensity of this chapter suggests that he may have come close to such a

mystical experience in the mountains of his beloved Tirol." See Moore, *Schrödinger*, 171. Schrödinger exhibited his admiration for the East in many places, for instance noting in another book that "in the Far East spiritual trends of tremendous consequence were started, connected with the names of Gautama Buddha (born about 560 B.C.), Laotse and his younger contemporary Confucius (born 551 B.C.)." See Schrödinger, *Nature and the Greeks and Science and Humanism*, 54–55. Schrödinger was extraordinarily well versed in Eastern philosophy for a nonspecialist. His biographer writes, "He did not merely read these books [on Hinduism] and make notes on them. He thought deeply about the teachings of the Hindu scriptures, reworked them into his own words, and ultimately came to believe them" (Moore, *Schrödinger*, 113). Moore's book is highly recommended as a masterly exposition of a magnificent mind. It's also worth noting that the way in which Schrödinger expresses his insight—that we are a "modification" of "infinite being"—is vintage Spinoza. We'll explore Spinoza in detail in Chapter 5.

60. Although Einstein rarely worked directly with the other geniuses of his time, he did co-author a paper with Wolfgang Pauli: "On the Non-existence of Regular Stationary Solutions of Relativistic Field Equations," *Annals of Mathematics* 44, no. 2 (1943): 131–137.

61. Jammer, *Einstein and Religion*, 236.

62. Wolfgang Pauli, for instance, took a long trip to India in 1952. See Meier, *Atom and Archetype*, xv. Werner Heisenberg also traveled to India: "In 1929 Heisenberg spent some time in India as the guest of the celebrated Indian poet Rabindranath Tagore [one of Einstein's favorites; more on Tagore in Chapter 5], with whom he had long conversations about science and Indian philosophy. This introduction to Indian thought brought Heisenberg great comfort, he told me" (Capra, *Uncommon Wisdom*, 43). And Niels Bohr visited India in 1960 as the guest of the prime minister: Ahmed Sherrif, "Where Brilliant Minds Met: The Chennai House That Left Neil [sic] Bohr Fascinated!," The Better India, Sept. 25, 2018, www.thebetterindia.com/160293 /chennai-scientists-neil-bohr-alladi-ramakrishnan/.

63. Einstein, "Chat About My Impressions in Japan," in *The Travel Diaries of Albert Einstein*, 246.

64. Einstein, *The Travel Diaries of Albert Einstein*, 246.

65. "A yearning for the Far East led me, in large part, to accept the invitation to Japan." From a letter Einstein wrote to Wilhelm Solf, then German ambassador to Japan, in *The Travel Diaries of Albert Einstein*, 253.

66. Einstein, *The Travel Diaries of Albert Einstein*, 117, 137, 143, 133.

67. Einstein, *The Travel Diaries of Albert Einstein*, 103, 139, 201, 143, 151, 171, 173, 179, 183. As is evident from these diary entries, Einstein was well aware that Shinto was a separate and older "natural religion" (173) that predated the introduction of Buddhism into Japan.

68. The downplaying and denial are discussed in this paragraph. For outright deletions of references to the Indian nondualist concept of Brahman, see the discussion of Einstein and Rabindranath Tagore in Chapter 5.

69. Einstein, *The Ultimate Quotable Einstein*, 453, 454.

70. Einstein to Norman Salit, Mar. 4, 1950, in Goldman, *Einstein's God*, 119 (see also Albert Einstein Archives document 61-226); Goldman, *Einstein's God*, 120.

71. Einstein, *Ideas and Opinions*, 41, 48 (he mentions Spinoza in the same breath as the Buddha—by no means an idle comparison, and one which we'll have occasion to explore in greater depth in Chapter 5); Jammer, *Einstein and Religion*, 236. Jammer continues: "On the contrary, he [Einstein] condemned 'the mystical trend of our time... [as] a symptom of weakness and confusion.'" The confusion, however, is Jammer's. This is clear when Einstein's quote is put back in its original context. The "mystical trend" Einstein had in mind was neither meditation nor Eastern mysticism, but a resurgent belief in the old fairy story that the soul survives the death of the body, which was being pushed by "spiritualist" groups headed by charlatans like Helena Blavatsky. Einstein's complete thought was as follows: "The mystical trend of our present time, especially evident in the enthusiastic growth of so-called theosophy and spiritualism, is to me a symptom of confusion and weakness. Since our inner experiences consist of reproductions and combinations of sensory impressions, the concept of a soul without a body seems to me to be empty and devoid of meaning" (*The Ultimate Quotable Einstein*, 446). Jammer's confusion is difficult to comprehend. He was well aware of Einstein's essay "Religion and Science," in which he praises Buddha and Buddhism. Jammer quotes liberally from this essay, even reproducing on p. 78 the exact quote where Einstein praises Buddhism—apparently oblivious to the fact that it blatantly contradicts his own reading of Einstein on p. 236. Einstein once wrote, "Whoever is careless of truth in small matters cannot be trusted in important affairs" (*The Ultimate Quotable Einstein*, 453). It's not hard to imagine what he'd think of Jammer's blatant misrepresentations. Jammer wasn't alone, however. Other experts on Einstein have made similar mistaken claims—the physicist Paul Halpern, for instance: "A critical difference between the beliefs of Einstein and Schrödinger was the latter's devotion to Eastern thought. None of the figures Einstein mentioned in his piece about religion [the 1930 essay in the *New York Times*] were from the Eastern tradition (he only briefly referred to Buddhism). He had little interest in any form of mysticism or spirituality. Schrödinger, on the other hand, had a deep sense that people share a common soul and that everything in nature is really a single entity." Halpern, *Einstein's Dice and Schrödinger's Cat*, 80.

72. On Gandhi's seventy-fifth birthday, Einstein gushed, "Generations to come... will scarce believe that such a one as this ever in flesh and blood walked upon this Earth." *The Ultimate Quotable Einstein*, 124.

73. Einstein's ardent pacifism and adherence to nonviolence—and the spiritual and philosophical underpinnings of these principles—will be explored in detail in Chapter 6.

74. As already mentioned, Einstein saw the quest for this sense of ultimate unity as "the one issue of true religion," and in Chapter 8 we'll explore his views on this topic in depth.

75. Hermanns, *Einstein and the Poet*, 108–109.

76. The condemnations of Einstein's cosmic religion by mainstream theologians included accusations from cardinals and bishops. This will be covered in detail in Chapter 5.

77. Dawkins, *The God Delusion*, 18.

78. "As long as you pray to God and ask him for something, you are not a religious man." Einstein, *The Ultimate Quotable Einstein*, 343.

79. "Everyone has been given an endowment that he must strive to develop in the service of mankind. This cannot be brought to completion through the threat of a God who will punish man for sin, but only by challenging the best in human nature." Einstein, *The Ultimate Quotable Einstein*, 334.

80. In answer to the question "What is your opinion regarding a 'savior'?" Einstein replied: "I can look at doctrinaire traditions only with a historical and psychological perspective; they have no other significance for me." *The Ultimate Quotable Einstein*, 324. Einstein also once contended, quite sacrilegiously, that "it is quite possible that we can do greater things than Jesus, for what is written in the Bible about him is poetically embellished." *The Ultimate Quotable Einstein*, 337.

81. "I have no faith in speculation about Heaven and Hell. I'm concerned with this time—here and now." Hermanns, *Einstein and the Poet*, 94.

82. "I cannot imagine a God who rewards and punishes the objects of his creation." Einstein, *The Ultimate Quotable Einstein*, 330.

83. "A man's ethical behavior should be based effectively on sympathy, education, and social relationships; no religious basis is necessary" (Einstein, *The Ultimate Quotable Einstein*, 328). "I believe that we have to…treat values and moral obligations as purely human problems" (*The Ultimate Quotable Einstein*, 338–339). "I consider ethics to be an exclusively human concern with no superhuman authority behind it" (*The Ultimate Quotable Einstein*, 341).

84. Dawkins's thoughts on the letter originally appeared on the Richard Dawkins foundation website, but the article no longer appears to be available. An archive of the original post is at https://web.archive.org/web/20140427025031/http://www.richarddawkins.net/news_articles/2012/8/15/albert-einstein-s-historic-1954-god-letter-handwritten-shortly-before-his-death.

85. Einstein, *The Ultimate Quotable Einstein*, 340.

86. Einstein, *The Ultimate Quotable Einstein*, 326.

87. Einstein, *The Ultimate Quotable Einstein*, 336.

88. Einstein to Guy Raner Jr., who had written Einstein a letter asking if it was true that he'd been converted to Catholicism by a Jesuit priest. A facsimile of the letter is at www.sothebys.com/en/buy/auction/2020/fine-books-and-manuscripts-including-the-eric-c-caren-collection/einstein-albert-typed-letter-signed-a-einstein-to.

89. Regarding the professorship in Prague: "Being a Jew was a disadvantage; being a nonbeliever who claimed *no* religion was a disqualifier. The empire required that all of its servants, including professors, be a member of some religion. On his official forms, Einstein had written that he had none." Einstein eventually relented and agreed to identify as "Mosaic." Isaacson, *Einstein*, 163.

90. "In his deposition, Einstein had given his religion as 'dissenter,' but in the divorce decree the clerk designated him 'Mosaic.'" Isaacson, *Einstein*, 243. Again, Einstein's attempt to exhibit his atheism was foiled.

91. Hoffmann and Dukas, *Albert Einstein: Creator and Rebel*, 144.

92. Einstein, *Ideas and Opinions*, 41.

93. Einstein had more books in his personal library by Bertrand Russell than any other single thinker. I counted at least sixteen in the Albert Einstein Archives in Jerusalem, and we know that his secretary, Helen Dukas, gave away many books

to friends and family after Einstein's death. So there very well might have been even more.

94. Bertrand Russell, *What Is an Agnostic?* (1953), https://scepsis.net/eng/articles /id_5.php.

95. Einstein, *The Ultimate Quotable Einstein*, 149–150.

96. *New York Times*, Mar. 19, 1940, in Einstein, *The Ultimate Quotable Einstein*, 149.

97. Beatrice Frohlich to Einstein, Dec. 12, 1952, in Jammer, *Einstein and Religion*, 121.

98. Einstein, *The Ultimate Quotable Einstein*, 341.

99. Einstein to Beatrice Frohlich, Dec. 17, 1952, in Jammer, *Einstein and Religion*, 121–122.

100. Einstein to an unidentified person, Aug. 7, 1941, in Jammer, *Einstein and Religion*, 97 (Albert Einstein Archives doc. 54-927).

101. Quoted in Jammer, *Einstein and Religion*, 97.

102. Quoted in Goldman, *Einstein's God*, 5.

103. Einstein, *Ideas and Opinions*, 50.

104. Einstein to Georg Count von Arco, Jan. 14, 1920, in *The Ultimate Quotable Einstein*, 322.

105. Commenting on this letter of Einstein's, Jammer crowed that "Einstein, as we see, was far from disputing the usefulness of religious education" (*Einstein and Religion*, 51). But this is hardly the correct conclusion to draw from Einstein's statements. And in his book on Einstein and God, McGrath quotes this saying only in part, giving the false impression that Einstein looked favorably upon belief in a personal God (*A Theory of Everything*, 146).

106. Einstein, *The Ultimate Quotable Einstein*, 333. Other founders of the new physics shared the sentiment. Max Planck, for instance—another Nobel laureate and the man usually credited alongside Einstein as the co-founder of the quantum revolution— also abhorred nihilism. In his book *Where Is Science Going?* he wrote that "those forms of religion which have a nihilist attitude to life are out of harmony with the scientific outlook and contradictory to its principles. All denial of life's value for itself and for its own sake is a denial of the world of human thought and, therefore, in the last analysis, a denial of the true foundation not only of science but also of religion." Quoted in Wilber, *Quantum Questions*, 151.

107. The phrases are from Richard Dawkins, in his discussion of the religious views of Einstein and other physicists, such as Stephen Hawking. The exact quote is: "Pantheism is sexed-up atheism. Deism is watered-down theism." See Dawkins, *The God Delusion*, 18.

3: World of Wonder

1. Einstein, *Ideas and Opinions*, 39.

2. Christianity is the quintessential example. Jesus's remarks on love are legion: "Love your enemies" (Matthew 5:44), "Love one another" (John 13:44), and "Love the Lord your God with all your heart and with all your soul and with all your mind" (Matthew 22:37) are just a few representative examples.

3. In contrast to Jesus's more cheerful outlook, living life in a perpetual state of "fear and trembling" is recommended by St. Paul in the New Testament (Philippians 2:12). The notion of *taqwa* in Islam also often places an emphasis on fear of Allah, although it has a variety of interpretations.

4. In Buddhism in particular, equanimity (Sanskrit upekṣā) with respect to both negative and positive emotions is usually seen as a highly prized goal and a crucial step on the path to enlightenment.

5. Einstein, "What I Believe." Also in *The Ultimate Quotable Einstein*, 331.

6. Socrates in Plato's *Theaetetus* 155d, trans. F. M. Cornford, quoted in Uzdavinys, *Philosophy as a Rite of Rebirth*, 2.

7. See, for instance, his eloquent explanation (Schopenhauer, *World as Will and Representation*, 1:200–207) of how solitude in nature can raise a sensitive soul to a feeling of sublime contemplation, a "state of elation" (201) where there is "complete emancipation from all willing and its cravings . . . only the state of pure contemplation" (203) and where "we are one with the world" (205). The scenario is discussed dispassionately, but like Schrödinger's similar "Vedantic vision," which we looked at in Chapter 2, it has all the hallmarks of autobiography. Schopenhauer even directly compares (205–206) his notion of the sublime state to the ultimate insight of the Vedas and Upanishads, "Thou art that."

8. Au-Young, *Lao Tze's Tao Teh King*, 113. This is one of several editions of the Tao Te Ching that Einstein owned; more on Einstein and Taoism in Chapter 5.

9. Hermanns, *Einstein and the Poet*, 108.

10. Hermanns, *Einstein and the Poet*, 14.

11. Hoffmann and Dukas, *Albert Einstein: Creator and Rebel*, 18.

12. The several preceding quotations are all from the epistemological essay "Physics and Reality" in Einstein, *Ideas and Opinions*, 320.

13. Einstein to Maurice Solovine, quoted in Goldman, *Einstein's God*, 24.

14. Hermanns, *Einstein and the Poet*, 7–8.

15. Gutfreund and Renn, *Einstein on Einstein*, 159.

16. Gutfreund and Renn, *Einstein on Einstein*, 159.

17. Gutfreund and Renn, *Einstein on Einstein*, 159.

18. Hermanns, paraphrasing Einstein's own words, in *Einstein and the Poet*, 8.

19. Hermanns, *Einstein and the Poet*, 90; Einstein, "The Religious Spirit of Science" (1934), in *Ideas and Opinions*, 43–44.

20. Hermanns, *Einstein and the Poet*, 70.

21. Hermanns, *Einstein and the Poet*, 63.

22. Einstein, *The Ultimate Quotable Einstein*, 331.

23. Einstein to Otto Juliusburger, Sept. 29, 1942, in *The Ultimate Quotable Einstein*, 56.

24. Gutfreund and Renn, *Einstein on Einstein*, 159.

25. Hermanns, *Einstein and the Poet*, 7. Einstein's lifelong secretary, Helen Dukas, agreed that the compass was a pivotal point in Einstein's life: "Einstein spoke often of the sense of wonder that came over him when he saw the compass. It was clearly a major event in his life." Dukas and Hoffmann, *Albert Einstein: The Human Side*, 19.

26. Gutfreund and Renn, *Einstein on Einstein*, 159.

27. All quotations here, including the words of Einstein's uncle Jakob, are as recalled by Einstein and quoted in Hermanns, *Einstein and the Poet*, 7.

28. Hoffmann and Dukas, *Albert Einstein: Creator and Rebel*, 24.

29. Aaron Bernstein's multivolume series, *From the Field of Natural Science*, published between 1853 and 1856.

30. Hermanns, *Einstein and the Poet*, 9.

31. Gutfreund and Renn, *Einstein on Einstein*, 157–158.

32. Gutfreund and Renn, *Einstein on Einstein*, 158.

33. Gutfreund and Renn, *Einstein on Einstein*, 158.

34. Gutfreund and Renn, *Einstein on Einstein*, 158.

35. Einstein, "What I Believe." Also in *The Ultimate Quotable Einstein*, 331.

36. Einstein, "What I Believe." Also in *The Ultimate Quotable Einstein*, 331.

37. Einstein, *Ideas and Opinions*, 41.

38. The story and quotes are given in Vallentin, *The Drama of Albert Einstein*, 10. The same story with slightly different phrasing is provided in Isaacson, *Einstein*, 262.

39. Einstein, *The Ultimate Quotable Einstein*, 7.

40. Einstein, *The Ultimate Quotable Einstein*, 533.

41. Einstein, *The Ultimate Quotable Einstein*, 501.

42. Einstein, *The Ultimate Quotable Einstein*, 534.

43. Bertrand Russell, preface to Einstein, *Einstein on Peace*, xiii.

44. Einstein, *The Ultimate Quotable Einstein*, 525.

45. Einstein, *The Ultimate Quotable Einstein*, 499.

46. Pais, *Subtle Is the Lord*, 7–8.

47. Attributed to Sir Arthur Eddington, in Isaacson, *Einstein: The Life of a Genius*, 57.

48. Pais, *Subtle is the Lord*, 11.

49. Einstein, *The Ultimate Quotable Einstein*, 524.

50. Einstein, "What I Believe."

51. Einstein, *The Ultimate Quotable Einstein*, 14–15.

52. Einstein even asserted that "setting an example is not the main means of influencing another, it is the *only* means." *The Ultimate Quotable Einstein*, 481.

53. Einstein, *The Ultimate Quotable Einstein*, 480.

54. Einstein, *Ideas and Opinions*, 46.

55. Einstein, *The Ultimate Quotable Einstein*, 168.

56. See, e.g., Keltner and Haidt, "Approaching Awe."

57. Einstein, "Why Do They Hate the Jews?," in *Out of My Later Years*, 250.

58. Hermanns, *Einstein and the Poet*, 90.

59. Hermanns, *Einstein and the Poet*, 138.

60. Moses brought down the famous Ten Commandments, meant as inviolable laws whose transgression was punishable with extreme measures (see Exodus 20). The Good Shepherd was a common way of conceiving of Jesus in the Gospels and was meant in a complimentary way: Jesus would be willing to lay down his life to protect his flock. But of course a shepherd implies sheep—that is, unquestioning followers. As for Mohammed, the very name he gave his religion, Islam, translates as "submission" or "total surrender."

61. Genesis 3:5, NKJV.

62. Genesis 3:6, NIV.

63. Genesis 3:22, NKJV.

64. Quoted in Singh, *Big Bang*, 1.

65. Hermanns, *Einstein and the Poet*, 72.

66. Quoted in Freeman, *The Closing of the Western Mind*, vii.

67. Quoted in Jones, *For the Glory of God*, 14.

68. For consummate analyses of how Buddhism's religiosity has been rationalized in the West, see, e.g., Thompson, *Why I Am Not a Buddhist*, and McMahan, *The Making of Buddhist Modernism*.

69. Einstein to Italian minister of justice Alfredo Rocco, Nov. 16, 1931, in *Einstein on Peace*, 154.

70. Einstein, "What I Believe."

71. Hermanns, *Einstein and the Poet*, 130.

72. Hermanns, *Einstein and the Poet*, 138.

73. Hermanns, *Einstein and the Poet*, 90–91.

74. Einstein, *The Ultimate Quotable Einstein*, 381.

75. Einstein, *The Ultimate Quotable Einstein*, 386.

76. Einstein, *The Ultimate Quotable Einstein*, 19.

77. Hermanns, *Einstein and the Poet*, 138.

78. Hermanns, *Einstein and the Poet*, 8.

79. Einstein, *The Ultimate Quotable Einstein*, 101.

80. Hermanns, *Einstein and the Poet*, 62.

81. Einstein, *The Ultimate Quotable Einstein*, 107.

82. Einstein, *The Ultimate Quotable Einstein*, 106.

83. "I am opposed to examinations—they only deter from the interest in studying. No more than two exams should be given throughout a student's [college] career. I would hold seminars, and if the young people are interested and listen, I would give them a diploma." Einstein, *The Ultimate Quotable Einstein*, 108.

84. "I am an opponent of every totalitarian system, whether Communistic or Christian. I believe in the versatility of the human mind and stand for its free development. And this is only possible when it is not bound to an institution, and when man relies on the regulative power of reason." Hermanns, *Einstein and the Poet*, 130.

85. "By academic freedom I understand the right to search for truth and to publish and teach what one holds to be true. This right also implies a duty: one must not conceal any part of what one has recognized to be true. It is evident that any restriction of academic freedom acts in such a way as to hamper the dissemination of knowledge among the people and thereby impedes rational judgment and action." Einstein, *The Ultimate Quotable Einstein*, 108.

86. Einstein, *The Ultimate Quotable Einstein*, 433.

87. Einstein, "What I Believe."

88. Einstein, "What I Believe."

4: The Hidden Harmony

1. Brian, *Einstein*, 61.

2. Banesh Hoffmann, physicist as well as friend to and biographer of Einstein, said: "Einstein once said something very interesting to me when we were trying to

think...he said, 'Ideas come from God.' Now he didn't believe in a personal God or anything like that. This was his metaphorical way of speaking. You cannot command the idea to come, it will come when it's good and ready. He put it in those terms: 'Ideas from God.'" Quoted in Brian, *Einstein*, 61.

3. Brian, *Einstein*, 61.

4. There are countless accounts of Einstein's *annus mirabilis*. Walter Isaacson does a very commendable job of making it all comprehensible to non-physicists (such as myself) in his bestselling biography, *Einstein*, 90–139. A very different, and in some ways much more intimate, account is given by Denis Brian, who focuses mostly on special relativity: *Einstein*, 60–68.

5. Einstein did get some crucial help from the more mathematically gifted mind of his good friend Michele Besso, and it's known that he bounced ideas off his first wife, Mileva Maric, who was trained in physics. But despite much scrutiny over the contributions of Besso and especially Maric, historians are now essentially in agreement that the work was Einstein's own.

6. Einstein in his third Rhodes Lecture at Oxford, May 23, 1931, quoted in Isaacson, *Einstein*, 352. Einstein was speaking here about his later quest for a unified field theory, but it's safe to say he felt the same way about his earlier work, especially revolutionary theories like special relativity.

7. I should confess at the outset that I won't be making much effort to distinguish between the "reliable" and "implausible" stories. There is actually a huge amount of material available—much more information than we have about the life of almost any other ancient thinker—but the problem is that the quality of the sources varies enormously, and they often contradict (or at least do not corroborate) one another. Yet after reading thousands of pages about Pythagoras and the Pythagoreans, I've come to the conclusion that every scholar sees what they want to see. Reasons (whether arbitrary or convincing) can always be found to discount certain pieces of evidence, dismiss certain stories, and doubt certain teachings. For centuries, scholars have doubted both the age and the reliability of the several surviving biographies of Pythagoras—and it's clear that all of them were written centuries after Pythagoras lived and died, so they've often been assumed to be legends and fairy tales. But a new consensus has emerged in recent decades: "It is now taken for granted that the biographies contain a great deal of early information about Pythagoras and his school, and much of the information is taken from older authorities whose work has since perished." Guthrie and Fideler, *Pythagorean Sourcebook*, 49.

8. Herodotus, *Histories*, IV.95, available at www.perseus.tufts.edu/hopper/text ?doc=Perseus:abo:tlg,0016,001:4:95.

9. Empedocles, fragment 129 in the DK (Diels-Kranz) system. The fragment does not refer to Pythagoras by name, but it's generally agreed among scholars that this is who is being praised. The fragments are available at https://en.wikisource.org/wiki /Fragments_of_Empedocles.

10. Iamblichus, *Life of Pythagoras*, §1, in Guthrie and Fideler, *Pythagorean Sourcebook*, 57.

11. Some say Sidon, instead, in nearby Phoenicia (more or less equivalent to modern-day Lebanon). See Iamblichus, *Life of Pythagoras*, §2, in Guthrie and Fideler, *Pythagorean Sourcebook*, 58.

12. Ferguson, *Music of Pythagoras*, 12, 14.

13. Guthrie, *A History of Greek Philosophy*, 1:31.

14. "Foreign traders congregated on Samos to exchange goods and, undoubtedly, cultural traditions between the Oriental world to the east—Chaldea, Syria, and Babylonia—and the Greek world to the west. This made Samos an important link between the early Greeks and the Persians during the era leading up to their wars in the following century." Joost-Gaugier, *Measuring Heaven*, 11. For the currents of Oriental thought and their huge impact on the Ionian philosophers, including Pythagoras, see Guthrie's classic *A History of Greek Philosophy*, 1:31–45. For the most detailed compilation and critical analysis of all existing ancient sources and fragments on Pythagoreanism, see Burkert's monumental *Lore and Science*. For a very readable translation of key texts attributed to Pythagoras and his followers, see the more accessible work of Guthrie and Fideler, *Pythagorean Sourcebook*. For the historical impact of Pythagoreanism on Western science and philosophy, see Ferguson, *Music of Pythagoras*, already mentioned. For a mind-boggling but much more controversial view, see Kingsley, *A Story Waiting to Pierce You*.

15. "As he was a youth devoted to learning, he left his country." Diogenes Laertius, *Life of Pythagoras*, §3, in Guthrie and Fideler, *Pythagorean Sourcebook*, 141. Likewise Porphyry, *Life of Pythagoras*, §1: "Pythagoras . . . early manifested studiousness," in Guthrie and Fideler, *Pythagorean Sourcebook*, 123.

16. If the eclectic list of travels and teachers that follows sounds implausible, remember that "the Greeks had always been interested in the traditions of older civilizations, with which they had been in contact for centuries," as one historian of science put it. "It is not by accident that the beginnings of Greek mathematics are credited to Thales and Pythagoras, both of whom were said to have lived in Egypt (and Pythagoras also in the East)." See Russo, *Forgotten Revolution*, 29. And just as ancient Greek culture was hardly a hermetically sealed container, neither was the nearby Near East, with which the ancient Greeks had so much contact. All kinds of ideas would have been diffusing in and out of these cultures (Babylon, Persia, and so on) from near and far. In other words: "Ancient cultures from the eastern Mediterranean to the Indian Ocean were shaped through a continuous interplay with one another, an interplay only dimly seen, which is the hidden map of ancient history. It is a map of caravan routes and sea voyages, of travels and commerce—and of their consequences . . . What [these routes] reveal is not a structure of parallel straight lines—one labeled 'Greece,' another 'Persia,' another 'India'—but a tangled web in which an element in one culture often leads to elements in others." See McEvilley, *The Shape of Ancient Thought*, 1.

17. Historians find reasons to doubt that Pythagoras studied with Thales (as they do everything), but in this case there's no good reason. Miletus was practically next door to Samos (about 35 km, as the crow flies), probably no more than a few days' travel even at that time. Both polities had been founded by the same wave of Greek migrants several centuries earlier, and both remained part of a common political union: the so-called Ionian League. And Pythagoras himself would have been no stranger to strange lands: his father, Mnesarchus, was by many accounts a merchant who traveled far and wide; in one account, he even took Pythagoras as far as southern Italy when he was a young boy. That would certainly help explain why Pythagoras migrated to Croton (in southern

Italy) later in life and founded his second school there—if he had firsthand knowledge of the region, it would obviously be a more attractive objective. More on this migration later. For the story of Thales falling down the well while contemplating the heavens, see Chapter 1.

18. For a discussion of Heraclitus's disdain, see Zhmud, *Pythagoras and the Early Pythagoreans*, 34.

19. The classicist Peter Kingsley points to "a well-attested and common—although until recently underrated—phenomenon during much of the first millennium BC: the phenomenon of specialized experts and craftsmen, both Greek and oriental, wandering across the Mediterranean and Near East in the areas between Italy in the West and Ionia, Mesopotamia, and central Asia in the East." This reality "gives the lie to the general assumption that, once Pythagoras had emigrated from Samos to southern Italy, Pythagoreanism effectively broke its ties with the East. This was, anyway, an extremely implausible assumption given the normal frequency and speed with which people as well as news and information travelled backwards and forwards between Italy and Sicily on the one hand, and Asia on the other." See Kingsley, *Ancient Philosophy, Mystery, and Magic*, 151–152. Elsewhere, Kingsley also points specifically to physical evidence in the form of religious artifacts which demonstrate cultural influence extending across Egypt, Carthage, and southern Italy, and notes that Carthaginians were both students and teachers among the Pythagorean communities. See Kingsley, "From Pythagoras to the Turba Philosophorum," 4, 5. Iamblichus likewise notes that "foreigners also joined the Pythagoreans." See his *Life of Pythagoras*, §34, in Guthrie and Fideler, *Pythagorean Sourcebook*, 115. More generally, it was long ago pointed out that "the populations of the cities [of Magna Graecia] were formed of very diverse elements." See Jardé, *The Formation of the Greek People*, 217.

20. For just one example, see Iamblichus, *Life of Pythagoras*, §30, in Guthrie and Fideler, *Pythagorean Sourcebook*, 99. But this central faith is attested by many ancient authors.

21. Diogenes, *Lives of the Eminent Philosophers*, §8.49.

22. Kingsley, *Ancient Philosophy, Mystery, and Magic*, 157.

23. Diogenes, *Lives of the Eminent Philosophers*, §8.49; Gaius Plinius Secundus (Pliny), *Natural History: A Selection*, trans. John F. Healy (Penguin Books, 1991), II.6.

24. Diogenes, *Lives of the Eminent Philosophers*, §8.85. According to another ancient writer, "Some insist that the earth is immovable; but the Pythagorean Philolaus says that it moves circularly around the central fire, in an oblique circle like the sun and moon"; quoted in Guthrie and Fideler, *Pythagorean Sourcebook*, 170. Stobaeus wrote, "Some Pythagoreans, among whom is Philolaus, suggest that the moon's resemblance to the earth consists in its surface being inhabited, like our earth"; quoted in Guthrie and Fideler, *Pythagorean Sourcebook*, 170. And the classicist Peter Kingsley notes, "So, for example, the Pythagorean Philolaus is held to have originated the idea of a celestial 'counter-earth' in the form of an invisible although inhabited planet, but he is also said to have propounded the mythical idea of the moon as an inhabited earth; and according to Heraclides Ponticus, one of Plato's companions, Pythagoreans spoke of the stars as heavenly 'earths,' doubtless inhabited as well." Kingsley, *Ancient Philosophy, Mystery, and Magic*, 92.

25. Diogenes, *Lives of the Eminent Philosophers*, §8.30.

26. Alcmaeon "was the first to identify the brain as the center of understanding and perception," according to Debernardi et al., "Alcmaeon of Croton," 247. "Alcmaeon's lifetime may have overlapped Pythagoras', and he was probably a Pythagorean himself. If not, he was close to them and clearly reflected Pythagorean thinking about opposites when he wrote, 'What preserves health is an equilibrium of the powers...health is a balanced mixture of opposites.'" Ferguson, *Music of Pythagoras*, 111. Alcmaeon wasn't the only outstanding doctor within Pythagoras's orbit. Democedes, famous as the ablest doctor of the ancient world, personal physician to King Darius of Persia, also studied medicine at Croton and was said to be a Pythagorean. See Iamblichus, *Life of Pythagoras*, §35, in Guthrie and Fideler, *Pythagorean Sourcebook*, 119.

27. Alcmaeon was a pioneer of dissection and experimentation, and his "most striking contribution was to establish the connection between the sense organs and the brain. Even the optic nerves and their chiasm (crossing) were clearly delineated. Going further he concluded that the brain was the organ of the mind, not only perceiving sensations but also responsible for thought and memory." Lyons and Petrucelli, *Medicine: An Illustrated History*, 192.

28. Kingsley, *Ancient Philosophy, Mystery, and Magic*, 146. There are many discussions in ancient literature of Archytas's mechanical ingenuity, and they go back at least as far as Aristotle (384–322 B.C.E.). Kingsley also adds (same page): "According to a number of reports Archytas' mechanical interests were shared by other Pythagoreans in Tarentum."

29. Iamblichus, *Life of Pythagoras*, §3, in Guthrie and Fideler, *Pythagorean Sourcebook*, 60. The Bible describes the prophet Elijah (c. 900–850 B.C.E., some four hundred years before Pythagoras) gathering hundreds of other religious practitioners there in order to prove the superiority of the Jewish God, Yahweh. Elijah builds an altar, offers sacrifices, and calls God down to Earth (in the form of fire) to consume the sacrifice. When he wins the spiritual contest, he has the rival prophets executed—a less than magnanimous gesture. See 1 Kings 18:18–40. The Bible also describes how he repaired an earlier altar there that was broken, implying a long history of religious use (1 Kings 18:30). The famous Catholic religious order the Carmelites would originate here, too, around the twelfth century. It remains a major religious site to this day, with Muslim mosques and Baha'i shrines still located on its slopes.

30. This remarkable story is told in Iamblichus's *Life of Pythagoras*. All quotes in this section are from Iamblichus, *Life of Pythagoras*, §3, in Guthrie and Fideler, *Pythagorean Sourcebook*, 60, unless otherwise noted.

31. Dillon and Hershbell, *Iamblichus*, §3, 43.

32. Iamblichus, *Life of Pythagoras*, §4, in Guthrie and Fideler, *Pythagorean Sourcebook*, 61.

33. Iamblichus, *Life of Pythagoras*, §5, in Guthrie and Fideler, *Pythagorean Sourcebook*, 62.

34. This wasn't just any old cavern, but the famous Idaean cave: a well-known center of initiation, the place where Zeus himself was supposed to have been raised, and a location where mortals could expect to come into contact with the divine. Even in Pythagoras's time, this was an ancient spiritual center, where religious practices had

already been ongoing for centuries—maybe even millennia. For further discussion, see Ustinova, *Caves and the Ancient Greek Mind*, 180.

35. Diogenes, *Lives of the Eminent Philosophers*, §8.41.

36. "According to Apuleius, Pythagoras is supposed to have reached even as far as India and been instructed by the Brahmans themselves." Schopenhauer, *Parerga and Paralipomena*, 1:39. He cites Apuleius's work *Florida* for this assertion, which I admit I haven't consulted myself.

37. "Now since, without doubt, Egyptian culture and religion came from India, as proven by the holiness of the cow together with a hundred other things, this explains Pythagoras' prescription of abstinence from an animal diet, namely the prohibition of slaughtering cows, as also the prescribed careful treatment of all animals; similarly his doctrine of metempsychosis, his white robes, his endless mystery-mongering, giving rise to symbolic aphorisms and even extending to mathematical theorems; furthermore the founding of a kind of priestly caste, with strict discipline and lots of ceremonial, the worship of the sun, and much else." Schopenhauer, *Parerga and Paralipomena*, 1:38. This seems to have been the first book by Schopenhauer that Einstein read, so he very likely would have come across this passage; more on this connection in Chapter 5.

38. See McEvilley, *The Shape of Ancient Thought*, ch. 7. "The Jainas sent missionaries to all parts of the world." The "Jains gravitated to the merchant class," and therefore "opportunities for travel were abundant. Religious mendicants and missionaries tend to follow trade, hitching rides along trade routes, and this seems to have been common in ancient India. The Jain ascetic who went back to Babylon with Alexander's army was not the first, nor the last" (all quotes from 204). McEvilley suggests that the Jains influenced the Orphics, themselves very active in Sicily and southern Italy. In turn, the connections between the Orphics and the Pythagoreans are abundant and agreed upon by almost all scholars. Hence a possible indirect route of influence from Jainism to Orphism to Pythagoreanism. This is only a theory, of course. *Proving* cultural influence two or three thousand years ago, without any written records, is pretty much out of the question. McEvilley, who mastered both ancient Greek and Sanskrit, spent twenty years conducting a monumental comparison of the philosophical schools of classical Greece and ancient India. In his magnum opus, *The Shape of Ancient Thought*, he shows definitively that many "diffusion channels" at the time would have allowed for ample cultural exchange. For diffusion channels in general, see McEvilley, *The Shape of Ancient Thought*, esp. 1–22. For a specific example with respect to the Pythagoreans: "The type of situation that would provide a concrete means of transmission is shown by the story of the physician Democedes of Croton [whom we met earlier], a contemporary of Pythagoras who spent years practicing medicine at the Persian court [in those days, next door to the empires of ancient India, which spread throughout modern-day Pakistan and Afghanistan] and then returned to Greece, no doubt full of foreign medical lore...In fact, Democedes returned specifically to Croton, where such ideas would have fed directly into the Pythagorean tradition." McEvilley is here discussing the specific possibility of the transmission of the idea of kundalini and other esoteric physiological theories of the mind and its development, but of course Democedes and others like him could in principle have transmitted all kinds of other cultural lore and even potentially specific spiritual practices. Quote from McEvilley, *The Shape of Ancient Thought*, 211.

39. Kingsley marshals an impressive array of evidence that really needs to be read to be believed. Correct or not, at the very least his theory explains many strange facts that have remained enigmatic for two and a half millennia, not least the identity of Abaris. As just one specific example, we know from biographies that Abaris gave Pythagoras a gift of great significance—a small golden dagger he had carried with him all the way from his homeland. Kingsley is certain this gift, so far unexplained, was in fact a *phurba*, an implement central to many rituals and meditation practices in Mongolian and Tibetan religion to this day. And if direct contact between Far Eastern shamans and the sage of Samos already sounds sufficiently incredible, it will be even harder to believe that an ancient painted portrait that appears to depict a Mongolian man was dug up in Tarentum, a Pythagorean stronghold not far from Croton (also discussed by Kingsley, but discovered independently by Italian archaeologists). See Kingsley, *A Story Waiting to Pierce You*, 72, 157.

40. Iamblichus, *Life of Pythagoras*, §17, in Guthrie and Fideler, *Pythagorean Sourcebook*, 74.

41. Porphyry, *Life of Pythagoras*, §41, in Guthrie and Fideler, *Pythagorean Sourcebook*, 131.

42. Iamblichus, *Life of Pythagoras*, §10, §32, in Guthrie and Fideler, *Pythagorean Sourcebook*, 68, 111.

43. "An inimitable quiet and serenity marked all his words and actions, soaring above all laughter, emulation, contention, or any other irregularity or eccentricity; his influence, at Samos, was that of some beneficent divinity." Iamblichus, *Life of Pythagoras*, §2, in Guthrie and Fideler, *Pythagorean Sourcebook*, 59.

44. Iamblichus, *Life of Pythagoras*, §29, in Guthrie and Fideler, *Pythagorean Sourcebook*, 98.

45. Porphyry, *Life of Pythagoras*, §40, in Guthrie and Fideler, *Pythagorean Sourcebook*, 131.

46. Iamblichus, *Life of Pythagoras*, §35, in Guthrie and Fideler, *Pythagorean Sourcebook*, 118.

47. "He taught that the soul is immortal, and that after death it transmigrates into other animated bodies." Porphyry, *Life of Pythagoras*, §19, in Guthrie and Fideler, *Pythagorean Sourcebook*, 126.

48. Porphyry, *Life of Pythagoras*, §19, in Guthrie and Fideler, *Pythagorean Sourcebook*, 126.

49. Iamblichus, *Life of Pythagoras*, §16, in Guthrie and Fideler, *Pythagorean Sourcebook*, 73.

50. Iamblichus, *Life of Pythagoras*, §17, in Guthrie and Fideler, *Pythagorean Sourcebook*, 75; Porphyry, *Life of Pythagoras* §46, in Guthrie and Fideler, *Pythagorean Sourcebook*, 132.

51. Guthrie and Fideler, *Pythagorean Sourcebook*, 46.

52. See Chapter 3.

53. Gutfreund and Renn, *Einstein on Einstein*, 159.

54. From an interview with Henry Russo in *The Tower*, quoted in Isaacson, *Einstein*, 17–18.

55. Einstein, *The Ultimate Quotable Einstein*, 323.

56. Einstein, *The Ultimate Quotable Einstein*, 340.

57. Hermanns, *Einstein and the Poet*, 50.

58. See Ferguson, *Music of Pythagoras*.

59. Ferguson, *Music of Pythagoras*, 314.

60. Guthrie and Fideler, *Pythagorean Sourcebook*, 45.

61. Schrödinger, *Nature and the Greeks and Science and Humanism*, 34, 36. Schrödinger even ranked the Pythagorean tradition as ultimately greater in importance than either the atomists or the Ionian school—usually considered the first scions, and true ancestors, of modern science (*Nature and the Greeks*, 50–51). The atomists usually include Democritus and Leucippus. The Ionian school refers to Thales, Anaximander, Anaximenes, Heraclitus, and Anaxagoras. However, since Thales was the first real teacher of Pythagoras (before the latter traveled on to Egypt and then immigrated to southern Italy), it makes little sense to consider them as belonging to different traditions. In fact, they are two branches of one and the same.

62. Heisenberg, *Physics and Philosophy*, 41–42.

63. Pauli, "Science and Western Thought" (1955), in *Writings on Physics and Philosophy*, 137–148.

64. Russell, *A History of Western Philosophy*, 29.

65. Russell, *A History of Western Philosophy*, 37.

66. "Philolaus and other possibly earlier Pythagoreans thought part of the apparent movement of the heavens was caused by the movements of the earth. Copernicus also referred to Hicetas [a Pythagorean who agreed that the Earth rotated on its axis] and Ecphantus of Syracuse [a Pythagorean who supported both a heliocentric model of the solar system, with a revolving Earth, as well as the idea of the Earth's rotation]." See Ferguson, *Music of Pythagoras*, 158. In summing up this influence, Ferguson writes: "The Pythagorean insight, from the sixth century B.C., that harmony and simple pattern expressed in numbers underlie nature clearly was for Copernicus a persuasively strong point in favor of his rearrangement of the cosmos" (245).

67. Quoted in Africa, "Copernicus' Relation to Aristarchus and Pythagoras," 403, 404.

68. Ferguson, *Music of Pythagoras*, 248–249. See also Thoren, *The Lord of Uraniborg*, 107. Kepler formulated his three laws of planetary motion based on Brahe's earlier astronomical observations. The third law explicitly relies on proportionality.

69. Galilei, *Dialogue Concerning the Two Chief World Systems*, 5.

70. Newton thought that "his theories of gravitation admirably supported the Pythagorean ideal of unity and simplicity." And in his fervor for the Pythagoreans, "Newton, in an extraordinary gesture, wrote that his own famous law of universal gravitation could be found in Pythagoras." Ferguson, *Music of Pythagoras*, 279.

71. Einstein, *Ideas and Opinions*, 42–43.

72. We've already discussed how Einstein spoke of Newton as possessed of the cosmic religious feeling and of how he thought Newton exemplified the reality that "in this materialistic age of ours the serious scientific workers are the only profoundly religious people" (see Einstein, *Ideas and Opinions*, 42–43). Einstein laid flowers on Newton's grave at Westminster Abbey on a trip to England (Isaacson, *Einstein*, 301), and even had a portrait of him hung in his Berlin study (Isaacson, *Einstein*, 248).

73. Einstein, *Ideas and Opinions*, 42.

74. Einstein, *The Ultimate Quotable Einstein*, 131.

75. Einstein, *The Ultimate Quotable Einstein*, 131.

76. Quoted in Ferguson, *Music of Pythagoras*, 288.

77. Ferguson, *Music of Pythagoras*, 252.

78. Quoted in Ferguson, *Music of Pythagoras*, 251.

79. Einstein's dismissal of revelation as a source of knowledge will be discussed in more detail in Chapter 8 and the Epilogue.

80. Hermanns, *Einstein and the Poet*, 68.

81. Jones, *For the Glory of God*, 3.

82. Whitehead's own words, quoted in Jones, *For the Glory of God*, 5.

83. See Jones, *For the Glory of God*, 7.

84. According to Jones, *For the Glory of God*, 7.

85. One list documents more than 160 miracles in the Bible, roughly one miracle every ten pages (depending on how tiny the print is in your Bible). See Spirit of Life Church, "How Many Miracles Are There in the Bible?," Feb. 28, 2019, www.spiritof lifeag.com/how-many-miracles-are-there-in-the-bible/.

86. If you're curious, consult the list already cited in the previous note.

87. Francis Collins, leader of the Human Genome Project and director of the National Institutes of Health, (in)famously claims there's no inherent conflict: John Horgan, "One of the World's Most Powerful Scientists Believes in Miracles," *Scientific American*, May 20, 2020, https://blogs.scientificamerican.com/cross-check /one-of-the-worlds-most-powerful-scientists-believes-in-miracles/.

88. Demon-donkeys: Numbers 22:21–35. God sending bears to maul unbelievers can, believe it or not, be found in 2 Kings 2:24. The rest of these risible stories are so well known I won't bother citing the specific passages.

89. See, for instance, Merzbach and Boyer, *A History of Mathematics*. They also highlight the role of Pythagoras in both transmitting and innovating on this material (44–47).

90. For instance, see Russo's incredible and somewhat heartbreaking *Forgotten Revolution*.

91. As a brilliantly witty article argued recently. But just because the persecution and thought policing might not have been as bad as we once thought, that still isn't much of a recommendation for this era. See Simon Winder, "The 'Dark Ages' Weren't as Dark as We Thought," *Literary Hub*, Apr. 24, 2019, https://lithub.com/the-dark -ages-werent-as-dark-as-we-thought/.

92. See, e.g., Russo, *Forgotten Revolution*. For a more specific look at mathematics and astronomy, see Neugebauer, *The Exact Sciences in Antiquity*.

93. See Gies and Gies, *Cathedral, Forge, and Waterwheel*, 81.

94. For details of the decline of curiosity and innovation in the Western world after the rise of Christianity, see Nixey, *The Darkening Age*, and Freeman, *The Closing of the Western Mind*.

95. Depending on which texts are included and the particular translation, Plato's complete works weigh in at around 450,000 words. Depending, again, on similar factors, the complete Bible (both Old and New Testaments) totals about 750,000 words.

96. Whitehead, *Process and Reality*, part II, ch. 1, §1, 39.

97. Aristotle, *Metaphysics*, 987ª29–31, in *Complete Works of Aristotle*, 2:1561.

98. Aristotle, *Metaphysics*, 987ª29–31, in *Complete Works of Aristotle*, 2:1561.

99. Proclus, *On the Theology of Plato*, book I, ch. V. See Taylor, *The Six Books of Proclus*, 64.

100. See Waterfield, *Plato of Athens*, 115.

101. So says the Roman statesman and philosopher Cicero in *De Oratore*, III 34.139. For details and discussion, see Huffman, *Archytas of Tarentum*, 7.

102. The idea of an esoteric oral tradition among the earliest Pythagoreans is an ancient one. As Walter Burkert, one of the foremost modern scholars of Pythagoreanism, put it: "Very often a distinction is made between a lower and a higher degree of Pythagorean wisdom, and this goes back at least as far as Timaeus [i.e., the fifth century B.C.E.]...there is also a distinction between *akousmatikoi* and *mathematikoi*. The *mathematici* are the 'genuine,' or truly 'philosophizing' members, whose goal is *akribeia*, so that they correspond to the 'esoteric' Pythagoreans." See Burkert, *Lore and Science*, 192.

103. See, for instance, *Phaedo*, 61d–e, where Plato puts Pythagorean oral teachings on the afterlife into the mouth of Socrates and explicitly says that the teaching he is conveying is akin to the doctrines of the Pythagorean Philolaus. Or *Gorgias*, 493a–c, where Socrates says, "Once I even heard one of the wise men say . . ." and then immediately refers to "a Sicilian, perhaps, or an Italian"—the strongholds of Pythagoreanism in Plato's day. See also *Republic*, 583b, and *Meno*, 81a. Quotations here are from *Plato: Complete Works*. For a discussion of why it's obvious that Pythagoreans are the ones being referred to, see the brilliant analysis by Kingsley, *Ancient Philosophy, Mystery, and Magic*, 90, 110–111.

104. Plato, *Letter XII* to Archytas of Tarentum: "I am overjoyed at receiving the treatises that have come from you and am filled with admiration for their author." In *Plato: Complete Works*, 1673.

105. On the creation of the cosmos, see the opening sections of Plato's *Timaeus*. On the immortality of the soul, see the Myth of Er, in Plato's *Republic*, 614–621. On the flight of the philosopher's soul, see Plato's *Phaedrus*, 246–250.

106. See the brilliant analysis in Kingsley, *Ancient Philosophy, Mystery, and Magic*, esp. 71–111.

107. According to a recent biographer, "*Timaeus* is one of Plato's richest and most impressive works....*Timaeus* constitutes an ingenious adaptation of the basic Pythagorean axioms that the universe is mathematical in structure and that mathematical principles are responsible for the beauty and goodness of the world." See Waterfield, *Plato of Athens*, 219–220. "Most scholars would date *Timaeus* among Plato's last works." See the introduction to the *Timaeus* in *Plato: Complete Works*, 1224.

108. *Timaeus*, 20a: "Take Timaeus here. He's from Locri, an Italian city under the rule of excellent laws. None of his compatriots outrank him in property or birth, and he has come to occupy positions of supreme authority and honor in his city. Moreover, he has, in my judgment, mastered the entire field of philosophy." The picture Plato paints of Timaeus as a kind of philosopher-king was almost certainly inspired by his time with Archytas, who played exactly this role in Tarentum at the time. *Plato: Complete Works*, 1228.

109. *Timaeus*, 29a, in *Plato: Complete Works*, 1235.

110. *Timaeus*, 30b, 32b, 30a, 32c, in *Plato: Complete Works*, 1236, 1237.

111. *Timaeus*, 53b–56c, in *Plato: Complete Works*, 1256–1259.

112. *Timaeus*, 57c, 56c, 69b, in *Plato: Complete Works*, 1260, 1259, 1270.

113. *Timaeus*, 35a–36b, in *Plato: Complete Works*, 1239.

114. *Timaeus*, 87, 88d, in *Plato: Complete Works*, 1287.

115. *Timaeus*, 90c, in *Plato: Complete Works*, 1289.

116. *Timaeus*, 90d, in *Plato: Complete Works*, 1289.

117. Kingsley, *Ancient Philosophy, Mystery, and Magic*, 131.

118. Galilei, *Dialogue*, vii. Einstein's foreword is also in Einstein, *Einstein on Politics*, 132–134.

119. Einstein, speech at Caltech, Jan. 25, 1932, in *Einstein on Peace*, 161; Einstein, "Newton's Mechanics and Its Influence on the Shaping of Theoretical Physics," in *CP* 15, doc. 503.

120. Einstein, "On the Method of Theoretical Physics," originally delivered as a lecture at Oxford on June 10, 1933, in *Ideas and Opinions*, 297. Einstein was well aware of the debt modern science owed to the ancient Greeks. Another book in his personal collection was Morris R. Cohen and I. E. Drabkin, *A Source Book in Greek Science* (McGraw-Hill, 1948), Albert Einstein Archives cat. 49D00187. This book exhaustively details the crucial contributions made by the Pythagoreans but also Democritus and Euclid, two of Einstein's other favorite Greek thinkers.

121. Einstein, "On the Generalized Theory of Gravitation," in *Ideas and Opinions*, 378–379.

122. Hermanns, *Einstein and the Poet*, 50.

123. Isaacson, *Einstein*, 440.

124. Einstein, *The Ultimate Quotable Einstein*, 385–386.

125. Iamblichus assures us that "anything better than this is not possible to find." Both quotes from Iamblichus, *Life of Pythagoras*, §33, in Guthrie and Fideler, *Pythagorean Sourcebook*, 114.

126. De Vogel, *Pythagoras and Early Pythagoreanism*, 197. Historian Kitty Ferguson makes an analogous point. "For a Pythagorean," she explains, "the path to knowledge about the universe and the path to reunion with the divine were one and the same path." Ferguson, *Music of Pythagoras*, 103.

127. Burkert, *Lore and Science*, 11.

128. Pauli, *Writings on Physics and Philosophy*, 140.

129. Heisenberg, *Physics and Philosophy*, 41; he's referring here to the Pythagorean school.

130. Russell, *A History of Western Philosophy*, 37.

131. Einstein, *Ideas and Opinions*, 43.

132. Einstein and Infeld, *Evolution of Physics*, 296.

133. Hermanns, *Einstein and the Poet*, 61.

134. Einstein and Infeld, *Evolution of Physics*, 213.

135. For this etymological association, see the discussion in Cameron, *The Pythagorean Background*, 26.

136. Cameron, *Pythagorean Background*, 26–27.

137. Cameron, *Pythagorean Background*, 26.

138. Cameron, *Pythagorean Background*, 27.

139. Einstein, "Remarks to the Essays Appearing in this Collective Volume," 684.

140. Einstein hardly ever mentions Pythagoras or his followers, but he does refer frequently to the story of proving the Pythagorean theorem as a young boy. In all my research, I discovered only two direct mentions of Pythagoras by name; both are quoted in the paragraphs below and turn out to be incredibly revealing.

141. Democritus was born either in Abdera, a few hundred kilometers away from Pythagoras's hometown on Samos, or possibly in the city of Miletus itself, where Pythagoras first went to study with Thales. Miletus was ground zero for this burgeoning of Greek philosophical thought. See Diogenes, *Lives of the Eminent Philosophers*, §9.34.

142. Diogenes, *Lives of the Eminent Philosophers*, §9.34–35.

143. Diogenes, *Lives of the Eminent Philosophers*, §9.46–49. See also Zhmud, *Pythagoras and the Early Pythagoreans*: "The first book about the Greek philosopher was Democritus' *Pythagoras*" (12). The book was lost, however, and unfortunately "we know nothing [of his book] beyond that it was full of admiration for Pythagoras" (28).

144. Taylor, *The Atomists*, 136; Smith, *Hippocrates*, 55–105. Smith points out that these remarkable stories of meetings between Democritus and Hippocrates are clearly later tellings, probably written sometime in the century before Christ. Though these are presented in the form of letters exchanged between Democritus and Hippocrates, obviously they are not actual letters. Nonetheless, the content and character of the stories are extraordinary, and even if they are no more than unverifiable legends, the letters at least tell us much about how Democritus was perceived by posterity in the centuries following his death. A fascinating discussion of the letters is also provided in Ustinova, *Divine Mania*, 341–342.

145. Diogenes, *Lives of the Eminent Philosophers*, §9.39.

146. Smith, *Hippocrates*, 57.

147. According to one of the pseudo-letters of Hippocrates, who was sent to cure Democritus's alleged madness: "It is not wholly madmen who want caves and quiet, but also those who scorn human affairs in their desire for freedom from perturbation.... And there, perhaps, Democritus, too, has been removed by learning. And so...he is judged mad for his love of solitude." Smith, *Hippocrates*, 63.

148. Einstein, *Ideas and Opinions*, 41.

149. Solovine, *Démocrite* (Einstein's personal copy, with a dedication from Solovine on the title page, is in the Albert Einstein Archives, catalogue #90C2285); Einstein to Maurice Solovine, Mar. 4, 1930, in Einstein, *Letters to Solovine*, 55.

150. My translation of Solovine's French: "La terre tout entière est ouverte à l'homme sage, car la patrie d'une âme élevée, c'est l'Univers." See Solovine, *Démocrite*, 144. Einstein's personal copy, with his handwritten marking of this passage, is in the Albert Einstein Archives, cat. #90C2285.

151. From a letter dated March 4, 1930; Einstein, *Letters to Solovine*, 55.

152. My translations of Solovine's French: "il fut un partisan zélé des Pythagoriciens." See Solovine, *Démocrite*, 3. I quote Solovine simply to show that Einstein must have known, beyond any doubt, that Democritus was a Pythagorean, but this is by no means Solovine's contention alone. Ancient authors like Diogenes long ago pointed

out Democritus's Pythagorean pedigree: "He is thought to have been an admirer of the Pythagoreans. And he himself mentions Pythagoras, expressing his admiration in his work entitled *Pythagoras*. He seemed to have taken all his ideas from him, and might even, if chronology did not stand in the way, be thought to have been his student." Diogenes adds, "According to Glaucus of Rhegium, his contemporary, Democritus certainly studied with one of the Pythagoreans. And Apollodorus of Cyzicus says that Democritus studied with Philolaus." Both quotes from Diogenes, *Lives of the Eminent Philosophers*, §9.38. Modern scholars agree. In his recent exhaustive study of the Pythagoreans, Leonid Zhmud states that "Democritus' own links with Pythagorean philosophy and science are undoubted...the influence of the Pythagoreans on Democritus was not limited to ethics. Aristotle more than once referred to their proximity in natural philosophy. Democritus' contacts with the Pythagoreans are evident in the scientific area too." Zhmud, *Pythagoras and the Early Pythagoreans*, 45.

153. My translation of Solovine's French: "en fréquentant parfois les lieux deserts et en séjournant parmi les sépulcres." See Solovine, *Démocrite*, 4.

154. Again, this is my translation of Solovine's original French: "Une conception surprenante par son audace et par les consequences incalculables qu'elle a eues dans l'histoire des science et de la philosophie est celle des Pythagoriciens. Ils proclament que «les éléments des nombres sont les éléments de toutes choses.» Les premiers ils ont reconnu le rôle que jouent les mathématiques dans la connaissance des choses." Solovine, *Démocrite*, xviii.

155. Quoted in Moszkowski, *Conversations with Einstein*, 182.

156. Bucky, *The Private Albert Einstein*, 31.

157. De Vogel, *Pythagoras and Early Pythagoreanism*, 197.

158. Einstein, *The Philosophy of Albert Einstein*, 53.

159. This comment came in a conversation about the nature of inspiration Einstein had with the Nobel Prize–winning poet Marie René August Alexis Léger. The conversation is recounted in Maurois, *Illusions*, 35; also quoted in Isaacson, *Einstein*, 549.

160. For an in-depth account of Einstein's creative process and the role of visual imagination in it, see Miller, *Imagery in Scientific Thought*. A now hard-to-find overview of how Einstein created special relativity, from the perspective of a Gestalt psychologist who knew Einstein well, is given in Max Wertheimer, *Productive Thinking* (Harper, 1959).

161. From the opening line of "How I Created the Theory of Relativity," a speech given on Dec. 14, 1922, during his visit to Japan. Detailed notes of the talk were published in Japanese in 1923, and later translated by Yoshimasa Ono and published in *Physics Today* in August 1982; the quote here is from p. 46 of that article.

162. Einstein, *The Ultimate Quotable Einstein*, 26.

163. As Einstein later explained the thought experiment, "If I pursue a beam of light with the velocity c (velocity of light in a vacuum), I should observe such a beam of light as a spatially oscillatory electromagnetic field at rest." But he soon realized that there was "no such thing" as *stationary* light, either in actual experience or in the equations of physics. Einstein claimed that he'd first had a vision of this kind at the age of sixteen: "After ten years of reflection such a principle resulted from a paradox upon which I had already hit at the age of sixteen." All quotes here are from Einstein, "Autobiographical Notes," 53.

164. For instance, general relativity, undeniably Einstein's greatest achievement, was likewise grounded in a visual thought experiment that Einstein described having in 1907, in which he once again imagined being an observer in outer space, confined within an elevator-like box and unable to determine whether the force he felt upon himself was due to gravitational attraction or an accelerating force. This thought experiment led Einstein to the seminal insight that acceleration and gravitation were indistinguishable—what Einstein later called the "equivalence principle." For details, see Miller, *Imagery in Scientific Thought*, 244. Many other examples are given throughout Miller's book. For a popular account highlighting many of Einstein's key visual thought experiments, see Isaacson, "The Light-Beam Rider."

165. Einstein, "On Science," in *Einstein on Cosmic Religion*, 97.

166. See, for instance, Isaacson, "The Light-Beam Rider." Almost every biography of Einstein also tells this story.

167. For a poetic translation and interpretation of these so-called Pyramid Texts, see Morrow, *The Dawning Moon of the Mind*. For a much more scholarly treatment, developed from a doctoral dissertation, see Naydler, *Shamanic Wisdom in the Pyramid Texts*.

168. Arthur Eddington's light deflection results of 1919 are the most famous confirmation of general relativity. But shortly after Einstein completed the theory in 1915, an opportunity arose to test the new theory against a tiny anomaly in the orbit of Mercury that had been known, but not explained, for decades. General relativity predicted the correct perihelion precession of Mercury with uncanny accuracy and was the theory's first great success. See Hoffmann and Dukas, *Albert Einstein: Creator and Rebel*, 124–125 for the details.

169. Hoffmann and Dukas, *Albert Einstein: Creator and Rebel*, 125.

170. Kepler, quoted in Ferguson, *Music of Pythagoras*, 275.

171. The words are spoken by Galileo's mouthpiece, Salviati (Italian for "sage"). See Galilei, *Dialogue*, 11. Einstein wrote a foreword to this modern edition of Galileo's book, and it's hard to imagine he missed the reference to the Pythagoreans, which occurs on the third page.

172. Russell, *A History of Western Philosophy*, 33.

173. Russell, *A History of Western Philosophy*, 33.

174. Hoffmann and Dukas, *Albert Einstein: Creator and Rebel*, 253.

175. Einstein, "Principles of Research," in *Ideas and Opinions*, 247.

176. Hermanns, *Einstein and the Poet*, 66.

177. Hermanns, *Einstein and the Poet*, 117.

178. Einstein, *Ideas and Opinions*, 297.

179. Einstein, *The Ultimate Quotable Einstein*, 446.

180. Einstein, *Ideas and Opinions*, 297.

181. Einstein, *The Ultimate Quotable Einstein*, 308.

182. Einstein, *Ideas and Opinions*, 300.

183. Quoted in Goldman, *Einstein's God*, 24.

184. Einstein, *Ideas and Opinions*, 300–301.

185. As famously asserted by Aristotle, *Metaphysics*, 987[a]29–31. See *Complete Works of Aristotle*, 2:1561.

186. Hermanns, *Einstein and the Poet*, 11–12.

187. Einstein, *Ideas and Opinions*, 300.

188. See Wigner, "The Unreasonable Effectiveness of Mathematics in the Natural Sciences."

189. Schopenhauer, "Fragments for the History of Philosophy," in *Parerga and Paralipomena*, 1:40, 39.

190. The forty-seventh saying from the Pythagorean Symbols or Maxims, in Guthrie and Fideler, *Pythagorean Sourcebook*, 161.

191. Schopenhauer, *Parerga and Paralipomena*, 1:39. Clement was writing many centuries after Pythagoras, so we might not think his conclusions carry much weight. But as Schopenhauer astutely notes, Clement is quoting an ancient source, and the quotation's "Dorian dialect points to its genuineness" (39). Doric was the dialect commonly spoken in southern Italy at the time the Pythagorean school flourished.

192. Lactantius, *The Divine Institutes*, I, 5, in Guthrie and Fideler, *Pythagorean Sourcebook*, 297.

193. From David Fideler's introduction to Guthrie and Fideler, *Pythagorean Sourcebook*, 33.

5: The Immanent Divine

1. Hermanns, *Einstein and the Poet*, 26; Einstein, *The Ultimate Quotable Einstein*, 153.

2. Part of the confusion is due to the fact that pantheism means many things to many different people. For a learned and lucid (albeit dense) look at Spinoza's particular brand, see Jaspers, *Spinoza*. For a classic overview of pantheism, see the lovely little book by Picton, *Pantheism*.

3. Both quotations are from Seidel, "Spinoza," 57.

4. These included Moses Maimonides, Hasdai Cresca, and Abraham ibn Ezra. For Spinoza's study of ancient Jewish authorities, and his departure from their teachings, see Kayser, *Spinoza*, 70; also Browne, *Blessed Spinoza*, 95–100.

5. Kayser, *Spinoza*, 53.

6. See Browne, *Blessed Spinoza*, 117. For more on van den Enden, see Nadler, *Spinoza*, ch. 5, esp. 120–133.

7. From the original writ of excommunication, quoted in Browne, *Blessed Spinoza*, 141.

8. For details of this tumultuous turning point in Spinoza's life, see Browne, *Blessed Spinoza*, ch. 6; Kayser, *Spinoza*, ch. 4; and Nadler, *Spinoza*, ch. 5.

9. Spinoza, *Treatise on the Emendation of the Intellect*, para. 1, in *Spinoza: Complete Works*, 3.

10. As reported by Balthasar Bekker in his 1691 book *The Enchanted World*; quoted in Kayser, *Spinoza*, 118.

11. Johannes Colerus and Jean-Maximilien Lucas, both quoted in Browne, *Blessed Spinoza*, 286.

12. "For a while it [Spinoza's *Short Treatise on God, Man, and the Understanding*] circulated secretly in manuscript among the members of Spinoza's circle in Amsterdam." Browne, *Blessed Spinoza*, 195–196.

13. Often considered the founder of mathematical physics. Among many other contributions, Huygens discovered the rings of Saturn and its moon Titan. He also invented the pendulum clock, the most accurate method of timekeeping for nearly three centuries.

14. From a letter written to Spinoza in 1673 by Johann Ludwig Fabritius, on behalf of Karl Ludwig, elector of Palatine, offering him a chair of philosophy at Heidelberg. Quoted in Nadler, *Spinoza*, 361.

15. Spinoza expressed these concerns clearly in his reply to Fabritius, dated Mar. 30, 1673, in which he declined the offer. See *Spinoza: Complete Works*, 887.

16. Huygens to his brother Constantijn, 1667, quoted in Nadler, *Spinoza*, 217; Leibniz, quoted in Nadler, *Spinoza*, 217.

17. The words are from Dirk Kerckerinck, who'd studied with Spinoza in Franciscus van den Enden's school when they were both young men. Quoted in Nadler, *Spinoza*, 217.

18. Nadler, *A Book Forged in Hell*, xi.

19. Quoted in Browne, *Blessed Spinoza*, 259.

20. For an in-depth exploration of the publication and immense importance of Spinoza's *Treatise*, see Nadler, *A Book Forged in Hell*. The quotes about Spinoza's book are from xi.

21. Einstein, *The Ultimate Quotable Einstein*, 340.

22. In the words of a petition authored by the orthodox in support of the appointment of a conservative pastor for the local church, instead of the liberal and progressive candidate who they thought was under Spinoza's influence. Quoted in Browne, *Blessed Spinoza*, 246.

23. Or as Spinoza put it, "The whole of Nature is but one only substance." Spinoza, *Short Treatise on God, Man, and His Well-Being*, part II, ch. XXII, in *Spinoza: Complete Works*, 94.

24. Spinoza, *Ethics*, part I, definitions 3 and 6, in *Spinoza: Complete Works*, 217.

25. From a letter Spinoza wrote to Jacob Ostens, probably sometime in 1671, in *Spinoza: Complete Works*, 880.

26. Spinoza, *Short Treatise on God, Man, and His Well-Being*, part II, ch. XXII, in *Spinoza: Complete Works*, 94.

27. See Dawkins, *The God Delusion*, 18.

28. Spinoza to Henry Oldenburg, in *Correspondence of Benedict Spinoza*, 28.

29. *Correspondence of Benedict Spinoza*, 28.

30. *Correspondence of Benedict Spinoza*, 28.

31. Spinoza, *Short Treatise on God, Man, and His Well-Being*, part II, ch. XVIII, in *Collected Works of Spinoza*, 1:85.

32. Spinoza, *Treatise on the Emendation of the Intellect*, para. 13, in *Spinoza: Complete Works*, 5–6.

33. Although he noted the similarities, Kant despised the pantheist perspective and even ranted against the notion of feeling oneself to be a part of Infinity. In a rancorous essay he wrote a few years before he died, he had this to say: "Now the person who broods on this [i.e., attaining a state of final tranquility and enlightenment] will fall into *mysticism*.... From this comes the monstrous system of Lao-kiun [i.e., Lao-tzu]

concerning the *highest good*, that it consists in *nothing*, i.e. in the consciousness of *feeling* oneself swallowed up in the abyss of the Godhead by flowing together with it, and hence by the annihilation of one's personality.... Hence the *pantheism* (of the Tibetans and other oriental peoples); and in consequence from its philosophical sublimation *Spinozism* is begotten." From Kant's essay "The End of All Things" (1794), in Kant, *Religion and Rational Theology*, 228.

34. Schopenhauer, *The World as Will and Representation*, 1:422.

35. Einstein, *Ideas and Opinions*, 48; Einstein, quoted in Viereck, *Glimpses of the Great*, 448.

36. As Spinoza is a seminal figure in the initiation of modern philosophy and the Enlightenment, scholars have left no stone unturned searching for his intellectual roots. They've yet to find any direct Eastern influence, and it's not hard to see why. We already saw in Chapter 2 how Eastern thought didn't really make its influence felt on Europe until the mid- to late 1700s, a century after Spinoza's death. There just weren't any translations of Eastern texts available back in the seventeenth century, and Spinoza was in no position to read anything in the original: there was no one to tutor him and no Eastern texts to read.

37. From a letter Spinoza wrote to John Hudde sometime in June 1666, in *Spinoza: Complete Works*, 857.

38. "One of Spinoza's own teachers, Menasseh ben Israel, was a keen Qabbalist. Indeed, Qabbalah was widely disseminated among Spanish Jewish scholars after the Expulsion." See *Spinoza: Complete Works*, 486n.

39. See, e.g., Popkin, "Spinoza, Neoplatonic Kabbalist?"

40. From Spinoza, *Theological-Political Treatise* (1670), ch. 9, in *Spinoza: Complete Works*, 486.

41. Various viewpoints are outlined in the classic study by McKeon, *The Philosophy of Spinoza*, 44–45. As far as I'm aware, to this day there is still no agreement on the matter. The main argument against any direct influence is that Bruno's books were not found in Spinoza's personal library after the latter's death. But while the *presence* of a book in a thinker's library can mean a considerable amount, its absence means little. Aside from the possibility of any given book being lost or given away, a man like Spinoza—himself hounded as a heretic and widely condemned by the authorities—probably wasn't dumb enough to leave the works of a man who was recently burned alive for his beliefs lying around in the open, even if he *had* owned or read them at some point. And given that Spinoza was close with various freethinkers, philosophers, and avant-garde scientists of his time, it's hard to imagine he hadn't at least heard about Bruno's ideas at second hand.

42. McKeon, *Philosophy of Spinoza*, 44.

43. Numerous Spinoza scholars and biographers (e.g., Christoph Sigwart, Richard Avenarius, Frederick Pollock, and Abraham Wolf) have argued that Spinoza must have read Bruno's *On Cause, Principle, and Unity*. For details and specific citations, see McKeon, *Philosophy of Spinoza*, 44.

44. Bruno, *Cause, Principle, and Unity*, 6.

45. Bruno, *Cause, Principle, and Unity*, 7.

46. Quote from Curley's introduction to the *Ethics* in *Collected Works of Spinoza*, 1:402.

47. Quote from Shirley's introduction to the *Ethics* in *Spinoza: Complete Works*, 214.

48. Browne, *Blessed Spinoza*, 219.

49. All quotes here are from Einstein's foreword in Runes, *Spinoza Dictionary*, v–vi. Einstein's own copy is still at the Albert Einstein Archives, cat. #52C00064.

50. Euclid, *The Thirteen Books of Euclid's Elements*, 1:2.

51. As discussed in detail in Chapter 4. The main source for Plato's time with the Pythagoreans is his "Seventh Letter." See *Plato: Complete Works*, 1646–1667.

52. As Heath has it: "I do not know of any reason for rejecting the evidence of the Scholia IV. Nos. 2 and 4 which say categorically that 'this Book' (Book IV) and 'the whole of the theorems' in it (including therefore Props. 10, 11) are discoveries of the Pythagoreans...while we have sufficient grounds for regarding the whole of the substance of Book II as Pythagorean." See Heath's appendix, "Excursus I: Pythagoras and the Pythagoreans," in Euclid, *The Thirteen Books of Euclid's Elements*, 414.

53. Einstein owned a book in which the author explains how "Pythagoras by tradition was a mathematician, and is supposed to have contributed with Thales much that appears in the first six books of Euclid"; Smith, *Man and His Gods*, 155. In itself, the simple fact that Einstein owned this book wouldn't be much to go on. But as it turns out, Einstein himself wrote the foreword and affirmed that, before providing his recommendation to the reader, he had read it "thoroughly and with intense interest." It's hard to imagine he missed this passage connecting two of his favorite thinkers. For the quote from Einstein's foreword, see Smith, *Man and His Gods*, ix.

54. See Jammer, *Einstein and Religion*, 48; Isaacson, *Einstein*, 388.

55. Isaacson, *Einstein*, 388–389.

56. Interview in the *Saturday Evening Post*, Oct. 26, 1929, quoted in Viereck, *Glimpses of the Great*, 447–448, and in part in Einstein, *The Ultimate Quotable Einstein*, 152; Einstein, letter of Nov. 23, 1951, in *The Ultimate Quotable Einstein*, 153.

57. Einstein, contribution to *Baruch Spinoza: Addresses and Messages*, 29. Einstein's copy of the book is in the Albert Einstein Archives, cat. #21V10974.

58. Schopenhauer, *The World as Will and Representation*, 2:87. "Altogether," Schopenhauer writes, "one might be surprised that pantheism did not gain complete victory over theism already in the seventeenth century; since the most original, beautiful and thorough European expositions of it (for compared to the *Upanishads* of the *Vedas* all of that is nothing) all became known during that period, namely through *Bruno*, *Malebranche*, *Spinoza*, and *Scotus Erigena*." "Sketch of a History of the Doctrine of the Ideal and the Real," in Schopenhauer, *Parerga and Paralipomena*, 1:9. He argues that the Pythagoreans were pantheists in the following essay (39).

59. Schopenhauer, *The World as Will and Representation*, 1:422.

60. Schopenhauer writes of Bruno: "Giordano Bruno (*Cause, Principle and Unity*, Dialogue 4) calls *transcendent* those predicates that are more universal than the distinction between corporeal and incorporeal substance and which, therefore, belong to substance in general; according to him, they refer to that common root in which the corporeal is one with the incorporeal and which is the true, original substance"; *Parerga and Paralipomena*, 1:76–77. And Schopenhauer directly compares Bruno's philosophy to that of Spinoza and also the *Bhagavad Gita* in the midst of a discussion of philosophical

systems that show us that death is "a false illusion"; *The World as Will and Representation*, 1:284.

61. The book is *Lichtstrahlen aus Giordano Bruno's Werken* (roughly, "Rays of Light from Giordano Bruno's Works"), ed. Ludwig Kuhlenbeck (H. W. T. Dieter, 1890). The handwritten inscription is in German and reads "Giordano Bruno / Das bist du gewesen / der bist du / 1929." The staff at the Einstein Archives in Jerusalem were unable to identify the unknown writer's handwriting for me. Einstein's personal copy of the book is at the Albert Einstein Archives, cat. #21V9043. Einstein's mention of Bruno's "deep religious feelings" shows that he must have been familiar with Bruno's pantheistic spiritual leanings, whether through reading this book or via some other unidentified source.

62. Hermanns, *Einstein and the Poet*, 60.

63. I've relied on several biographies of Bruno for this section of the book. The most readable is White, *Pope and the Heretic*. I also recommend Rowland, *Giordano Bruno*, and the dense scholarly treatment given in the classic by Yates, *Giordano Bruno*.

64. Bruno, *On the Infinite*, 79, 32.

65. Bruno, *On the Infinite*, 33.

66. Bruno, *On the Infinite*, 8.

67. Bruno, *On the Infinite*, 7.

68. Bruno, *On the Infinite*, 7.

69. Bruno, *On the Infinite*, 33.

70. Bruno, *On the Infinite*, 8.

71. According to Bruno biographer Michael White, "the charges" against him involved "a total of eight counts of heresy. These included his belief that the transubstantiation of bread into flesh and wine into blood was a falsehood [and] that the virgin birth was impossible." See White, *Pope and the Heretic*, 3.

72. Jole Shackelford, "Myth 7: That Giordano Bruno Was the First Martyr of Modern Science," in Numbers, *Galileo Goes to Jail*, 66; White, *Pope and the Heretic*, 3. As Shackelford writes: "The fact remains that cosmological matters, notably the plurality of worlds, were an identifiable concern all along and appear in the summary document"—that is, the document that summarizes the charges brought against Bruno by the Roman Inquisition. "Bruno was repeatedly questioned on these matters, and he apparently refused to recant them at the end. So, Bruno probably was burned alive for resolutely maintaining a series of heresies, among which his teaching of the plurality of worlds was prominent but by no means singular" (66).

73. This account of Bruno's torture and execution is based on Fitzgerald, *Discourse on Civility and Barbarity*, 239. There are many other more or less detailed accounts of what happened on that dark day in Rome.

74. Schopenhauer, *The World as Will and Representation*, 1:423.

75. Notably, most ancient authors tell us that Pythagoras too was murdered by an angry mob. Ancient biographers differ about how he met his end, either by being burned alive (when a mob burned down his school), or by being chased down in a field and killed with a weapon, or by voluntarily starving himself to death in seclusion after the mob had destroyed his school and murdered his followers. And as we discussed earlier in this chapter, Spinoza only narrowly dodged an assassin's dagger.

76. All quotes from *New York Times*, Nov. 30, 1930. The highlights of this episode are provided in Jammer, *Einstein and Religion*, 82, and Clark, *Einstein: The Life and Times*, 425–426. But the more detailed coverage of Sheen's rantings in the original *New York Times* article is well worth reading as an example of the vitriolic narrow-mindedness of Einstein's contemporary critics.

77. There were many unfair attacks on Einstein and relativity, but maybe the most egregious example is that of the pro-Nazi, Nobel Prize–winning physicist Philip Lenard, who waged a lifelong war against Einstein and his "Jewish physics." See Philip Ball's excellent "How 2 Pro-Nazi Nobelists Attacked Einstein's 'Jewish Science,'" *Scientific American*, Feb. 13, 2015 (the article is an excerpt from Ball, *Serving the Reich*).

78. For instance, in 1932 Einstein made a public speech honoring Spinoza in the highest possible terms, some of which has already been quoted in this chapter. See Einstein, contribution to *Baruch Spinoza*. In 1934 Einstein published his first book of personal thoughts and reflections on various social, scientific, and philosophical issues, titled *Mein Weltbild* (published in English in 1935 as *The World as I See It*). This volume contained a brief essay, "The Religious Spirit of Science," which clearly confirms Einstein's pantheistic and Pythagorean view of the cosmos as an integrated and well-ordered totality, comprehensible with mathematics and worthy of reverence. The essay is reproduced in Einstein, *Ideas and Opinions*, 43–44. And in 1936 Einstein received a charmingly ingenuous letter from a young Christian girl named Phyllis Wright in New York, asking him whether scientists prayed, and if so, what for. Einstein's reply focused on his Pythagorean faith ("ultimately the belief in the existence of fundamental all-embracing laws also rests on a sort of faith") and on his pantheistic feelings that the divine filled the physical world ("everyone who is seriously engaged in the pursuit of science becomes convinced that the laws of nature manifest the existence of a spirit vastly superior to that of men…the pursuit of science leads therefore to a religious feeling of a special kind, which differs essentially from the religiosity of more naïve people"). From Einstein's reply to P. Wright, Jan. 24, 1936, Albert Einstein Archives doc. 52-337, also quoted in Jammer, *Einstein and Religion*, 92–93.

79. For instance, we've already quoted a letter that Einstein wrote late in life (1946), where he still maintained that "Spinoza is one of the deepest and purest souls our Jewish people has produced." Einstein, *The Ultimate Quotable Einstein*, 153.

80. All quotes in this paragraph are from Einstein's essay "Science and Religion," in *Ideas and Opinions*, 47–53.

81. Comments made by a Catholic priest who wished to remain anonymous. Quoted in Jammer, *Einstein and Religion*, 98.

82. *Time*, Sept. 23, 1940, at https://content.time.com/time/subscriber/article/0 ,33009,801992-1,00.html; also quoted in Sayen, *Einstein in America*, 158.

83. Letter from a historian from New Jersey to Einstein, Nov. 14, 1940, Albert Einstein Archives doc. 40-339. Also quoted in Jammer, *Einstein and Religion*, 104.

84. Letter from Einstein to an unidentified person, Aug. 7, 1941, Albert Einstein Archives doc. 54-927. Also quoted in Jammer, *Einstein and Religion*, 97.

85. A.C.G. to Einstein, Sept. 11, 1940, Albert Einstein Archives doc. 40-247. Also quoted in Jammer, *Einstein and Religion*, 103.

86. Dr. Burris Jenkins of the Community Church, quoted in Jammer, *Einstein and Religion*, 99.

87. Einstein to Raymond Benenson, Jan. 31, 1946, Albert Einstein Archives doc. 56-505, also in *The Ultimate Quotable Einstein*, 396–397; Hermanns, *Einstein and the Poet*, 82.

88. Einstein to an admirer, Mar. 22, 1954 (just a year before he died), in *The Ultimate Quotable Einstein*, 342–343. He explained himself in almost identical words in a letter written a few weeks later: "I don't try to imagine a God," he said. "It suffices to stand in awe of the structure of the world." Einstein to S. Flesch, Apr. 16, 1954, in *The Ultimate Quotable Einstein*, 343.

89. Hermanns, *Einstein and the Poet*, 63–64.

90. Hermanns, *Einstein and the Poet*, 128.

91. Hermanns, *Einstein and the Poet*, 61, 128.

92. Hermanns, *Einstein and the Poet*, 69.

93. Einstein describes reading the book in a letter to Marcel Grossman, Sept. 6, 1901, quoted in Howard, "A Peek Behind the Veil of Maya," 93. Einstein didn't read Spinoza's *Ethics* until two years or so later (probably in 1903), as part of the self-curated curriculum of great books read by the so-called Olympia Academy he formed with a few close friends. Of course, evidence regarding what Einstein read, and exactly when, is sparse. The letter to Grossman indicates only the latest possible date at which Einstein began reading Schopenhauer. Given that he read Kant's *Critique of Pure Reason* at thirteen, and that Schopenhauer was Kant's greatest (indirect) disciple, Einstein might well have been reading Schopenhauer earlier than 1901.

94. As Schopenhauer says: "Altogether one might be surprised that pantheism did not gain complete victory over theism already in the seventeenth century; since the most original, beautiful and thorough European expositions of it (for compared to the *Upanishads* of the *Vedas* all of that is nothing) all became known during that period, namely through Bruno, Malebranche, Spinoza, and Scotus Erigena." See Schopenhauer, *Parerga and Paralipomena*, 1:9. Since we can count on one hand the historical figures Einstein praised for their profound religious feelings (as discussed throughout the book, these people are the Buddha, Giordano Bruno, Democritus, Francis of Assisi, and Spinoza), it's hard to believe it's just a coincidence that two of them appear as paragons of pantheism on *page three* of the first book Einstein read by Arthur Schopenhauer, where they're directly linked to the nondualist vision of Advaita Vedanta. Einstein no doubt had great admiration for the spiritual attainments of many other figures (including living exemplars like Mahatma Gandhi), but the five listed above are the only ones I'm aware of where there is a direct indication, in Einstein's own words, that he saw them as enlightened or saintly.

95. Deussen, *The Philosophy of the Upanishads*, 39.

96. As discussed in detail in Chapter 2.

97. Schopenhauer, *The World as Will and Representation*, 1:355.

98. Schopenhauer, *The World as Will and Representation*, 2:463, 1:411.

99. Schopenhauer, *The World as Will and Representation*, 1:411.

100. Einstein to a Hindu correspondent, Mar. 24, 1950, in *Einstein on Peace*, 525. Much more will be said about Einstein's relationship to Gandhi in Chapter 6.

101. Gandhi, *Wit and Wisdom*, 27.

102. Gandhi, *Wit and Wisdom*, 28.

103. Gandhi, *Wit and Wisdom*, 29.

104. Jordens, *Gandhi's Religion*, 123.

105. Jordens, *Gandhi's Religion*, 123.

106. See Browne, *The World's Great Scriptures*. The author was the same man who wrote the beautiful biography *Blessed Spinoza*, which was also in Einstein's possession.

107. Browne, *The World's Great Scriptures*, 71.

108. See the translated selections from the Mundaka Upanishad in Browne, *The World's Great Scriptures*. All quotes from 80.

109. Tagore, *The Religion of Man*, 63.

110. Tagore, *The Religion of Man*, 64.

111. Pais, *Einstein Lived Here*, 102.

112. It should be noted that Einstein was dissatisfied with the transcripts of these talks. "My conversation with Tagore was rather unsuccessful because of difficulties in communication and should, of course, never have been published," he wrote later that year (Einstein to Romain Rolland, Oct. 10, 1930, in *Einstein on Peace*, 112). But we have to take what we can get; any record at all of a meeting between these two titans, however limited, is of great value. And while the specifics of the conversation might be a bit garbled, the only important point for our purposes is that Tagore definitely told Einstein about his "Religion of Man," and definitely linked his ideas to the nondualism of Advaita and the Upanishadic reverence for Brahman.

113. From "Note on the Nature of Reality," a conversation between Tagore and Einstein that took place on July 14, 1930, in Tagore, *The Religion of Man*, 221.

114. Tagore, *The Religion of Man*, 223.

115. I found ten books by Tagore in Einstein's personal library, now housed at the Albert Einstein Archives in Jerusalem. And it's possible that others were either lost or given away.

116. From Einstein's contribution to *The Golden Book of Tagore*, an homage to the poet published in 1931, quoted in Pais, *Einstein Lived Here*, 108.

117. Einstein, *Ideas and Opinions*, 41.

118. Einstein, *Ideas and Opinions*, 41.

119. Einstein, *Ideas and Opinions*, 41.

120. As noted in Schrödinger, "What Is Real?" (1960), §3, in Schrödinger, *My View of the World*, 84.

121. Bohr claimed that modern physicists were now facing "those kinds of epistemological problems with which already thinkers like the Buddha and Lao Tzu have been confronted." Quoted in Capra, *Tao of Physics*, 18.

122. Quotation from Pauli to Jung, Feb. 27, 1952, in Meier, *Atom and Archetype*, 74. For Pauli's interest in Taoism, see his *Writings on Physics and Philosophy*, 138–139. See also Pauli's repeated references to Lao-tzu, the Tao, and the Tao Te Ching in his letters to Carl Jung over many years: e.g., Pauli to Jung, May 24, 1934, in Meier, *Atom and Archetype*, 27; Pauli to Jung, Feb. 27, 1952 (*Atom and Archetype*, 74–75); and Pauli to Jung's longtime assistant Aniela Jaffé, Dec. 12, 1951 (*Atom and Archetype*, 73).

123. Einstein, "The Goal of Human Existence," in *Out of My Later Years*, 260. Einstein owned *Kung-tse: Leben und Werk* (Confucius: Life and Work) by Richard Wilhelm.

Additionally, Lewis Browne's *The World's Great Scriptures*, which Einstein owned, contains substantial selections from the *Analects* of Confucius in English translation. That volume also contains substantial selections from the writings attributed to Chuang Tzu (328–358). It's telling that Browne titled the very first section of his book "All Things Are One" (328). And Einstein's copies of the Tao Te Ching included translations into both English and German. Besides two different editions of Richard Wilhelm's German translation, Einstein also owned another German version by F. Fiedler from 1899, as well as an English version entitled *Lao Tze's Tao Teh King: The Bible of Taoism*, translated by Sum Nung Au-Young, from 1938. Browne's *The World's Great Scriptures* also contains an almost-complete rendition of Lin Yutang's translation of the Tao Te Ching, reproduced from his 1942 book *The Wisdom of China and India*.

124. Einstein's personal copy, with Au-Young's handwritten dedication, is in the Albert Einstein Archives in Jerusalem, cat. #90C817.

125. The book is *Laotse: Tao Te King*, translated into German by F. Fiedler and published by Paul Steegeman Verlag in 1899. Einstein marked eighteen of the traditional eighty-one sections. Translations of the Tao Te Ching are remarkably variable, and so consulting multiple versions is a must, but for the sake of brevity I've chosen a single English translation for each relevant passage. Using the traditional Wang Pi ordering of the text, the eighteen sections Einstein marked are as follows: 28, 29, 33, 36, 37, 43, 44, 45, 47, 49, 50, 53, 56, 63, 70, 71, 76, and 81.

126. The Albert Einstein Archives in Israel maintains a list of all books with handwritten markings within them. I looked through them all and actually found markings in a few additional books that had not been included on the list.

127. Besides the Tao Te Ching, Wilhelm also translated the *Analects* of Confucius (which Einstein also owned), the famous I Ching (Book of Changes), and a Taoist meditation manual entitled *The Secret of the Golden Flower*, for which his friend Carl Jung provided a lengthy commentary. Much more will be said about Einstein and Carl Jung in Chapter 9.

128. I am quoting, of course, from an English translation of Wilhelm's German edition. Lao Tzu, *Tao Te Ching*, 65.

129. Lao Tzu, *Tao Te Ching*, 66.

130. Lao Tzu, *Tao Te Ching*, 115.

131. Lao Tzu, *Tao Te Ching*, 19.

132. Au-Young, *Lao Tze's Tao Teh King*, §34, 63.

133. Hermanns, *Einstein and the Poet*, 55.

134. The classic case of this "argument from design" is William Paley's famous thought experiment, in which he asks us to imagine walking through an open field and coming across a watch lying on the ground. Whereas something as natural as a rock needed no explaining, Paley thought that anything as purposeful as a pocket watch obviously implied a watchmaker (i.e., a Designer). For Paley, this was all the more true when we looked at the awe-inspiring complexity of living things—and therefore living beings, too, must have had a Designer.

135. Walter Isaacson, "Einstein & Faith," *Time*, Apr. 5, 2007, http://content.time .com/time/subscriber/article/0,33009,1607298-1,00.html.

136. Einstein, *The Ultimate Quotable Einstein*, 536.

137. For the misquotations in Isaacson's lecture at the Aspen Institute, see "Walter Isaacson—Einstein's God," YouTube, posted by FORA.tv, June 25, 2007, www.youtube .com/watch?v=S7r57oCT2cU, at 2:15–2:47.

138. Einstein, *The Ultimate Quotable Einstein*, 334.

139. Elsewhere Einstein made the same point in more detail. "First, you must have faith in an external world independent of you; then you must have faith in your ability to perceive it, and finally you must try to explain it by means of concepts or mathematical constructions." Hermanns, *Einstein and the Poet*, 139. See also the discussion in Chapter 4.

140. Quoted in Goldman, *Einstein's God*, 33.

141. Hermanns, *Einstein and the Poet*, 65.

142. Einstein, *The Ultimate Quotable Einstein*, 341.

143. Hermanns, *Einstein and the Poet*, 89.

144. Einstein, *The Ultimate Quotable Einstein*, 324.

145. Einstein, *The Ultimate Quotable Einstein*, 325.

146. Hermanns, *Einstein and the Poet*, 71.

147. Elkana, "Einstein and God," 39.

148. So says Banesh Hoffmann, for instance, Einstein's close friend and biographer. "He didn't believe in a personal God or anything like that. This was his metaphorical way of speaking"; quoted in Brian, *Einstein*, 61. Leopold Infeld, another close friend and collaborator of Einstein's, concurred. "When Einstein speaks of God he means the intrinsic relationships and logical simplicity of the laws of nature"; quoted in Kuznetsov, *Albert Einstein*, 60.

149. Einstein, *The Ultimate Quotable Einstein*, 342.

150. Einstein, *The Ultimate Quotable Einstein*, 324.

151. Au-Young, *Lao Tze's Tao Teh King*, 80. Einstein marked section 47 in his German edition by Fiedler, but here I've reproduced an English translation from the edition by Au-Young, which Einstein also owned and probably read.

152. From Einstein's interview with Georg Viereck, *Saturday Evening Post*, Oct. 26, 1929; quote from 110. The full, unedited interview is reprinted in Viereck, *Glimpses of the Great*. Einstein was referring here specifically to the electromagnetic force, but as we've seen, he was adamant that all the natural forces were ultimately derivable from a single arch-force. And in fact even this specific quote is taken from a discussion about Einstein's search for a unification of the forces of gravity and electromagnetism, part of his quest for unity in nature.

153. Einstein, *The Ultimate Quotable Einstein*, 324.

154. Schopenhauer, *Parerga and Paralipomena*, 1:68.

155. Hermanns, *Einstein and the Poet*, 63.

156. Hermanns, *Einstein and the Poet*, 61.

6: A Higher Calling

1. Einstein, *The Philosophy of Albert Einstein*, 58.

2. Hermanns, *Einstein and the Poet*, 93.

3. Hermanns, *Einstein and the Poet*, 133.

4. Hermanns, *Einstein and the Poet*, 141.

5. Schopenhauer, *World as Will and Representation*, 2:600.

6. Hermanns, *Einstein and the Poet*, 66.

7. From Otto Nathan's introduction to Einstein, *Einstein on Peace*, v.

8. From Otto Nathan's introduction to Einstein, *Einstein on Peace*, vii, v.

9. Originally published in 1960, it was the first collection of Einstein's writings that appeared after his death.

10. Einstein, *Einstein on Peace*, vi.

11. Einstein, *Einstein on Peace*, vii.

12. Einstein, *Einstein on Peace*, vii.

13. Einstein, *Einstein on Peace*, vii.

14. Einstein, *Einstein on Peace*, vii.

15. Einstein, *The Ultimate Quotable Einstein,* 247.

16. Einstein, *The Ultimate Quotable Einstein,* 279.

17. From a message to the General Conference of the Methodist Church, Apr. 27, 1952, in Einstein, *Einstein on Peace*, 566.

18. The story is told and the quote given in Isaacson, *Einstein*, 21. However, no original source is provided.

19. Einstein, *The Ultimate Quotable Einstein*, 104.

20. "Einstein's contempt for Germany's authoritarian schools and militarist atmosphere made him want to renounce his citizenship in that country. This was reinforced by Jost Winteler, who disdained all forms of nationalism and instilled in Einstein the belief that people should consider themselves citizens of the world." See Isaacson, *Einstein*, 29.

21. Einstein, "What I Believe."

22. See the "Manifesto to the Europeans," which Einstein signed in Oct. 1914, shortly after the outbreak of war. Reproduced, with excellent background and commentary, in Einstein, *Einstein on Politics*, 64–67.

23. Einstein to the king of Belgium, July 14, 1933, during the rise of the Nazis to power. In *Einstein on Peace*, 227–228.

24. From a letter dated Mar. 20, 1951, in reply to someone who brought Einstein's attention to a conscientious objector who had been sentenced to ten years in prison in Kansas. In Einstein, *Einstein on Peace*, 542–543.

25. From a message to the War Resisters' International, Aug. 10, 1953, in Einstein, *Einstein on Peace*, 593.

26. Albert Einstein, "The Road to Peace," *New York Times Magazine*, Nov. 22, 1931, in *The Ultimate Quotable Einstein*, 251.

27. From a United Nations Radio interview Einstein gave on June 16, 1950, recorded at his home in Princeton, in *The Ultimate Quotable Einstein*, 124.

28. From an interview Einstein gave to *Survey Graphic* magazine, published in Aug. 1935, in *The Ultimate Quotable Einstein*, 123.

29. Hermanns, *Einstein and the Poet*, 52–53.

30. Hermanns, *Einstein and the Poet*, 53.

31. Einstein to Seiei Shinohara, Feb. 22, 1953, in *Einstein on Politics*, 490–491.

32. Einstein to Seiei Shinohara, June 23, 1953, in *Einstein on Politics*, 491–492.

33. Einstein, statement made on Feb. 11, 1948, at a memorial service held in Washington, D.C., following Gandhi's assassination, in *Einstein on Peace*, 467.

34. Einstein, *The Ultimate Quotable Einstein*, 413; Alice Calaprice, "Einstein's Last Musings," *Princeton University Library Chronicle* 65, no. 1 (2003): 68. On the family pets, see also Isaacson, *Einstein*, 438.

35. Einstein, *The Ultimate Quotable Einstein*, 415.

36. Einstein, *The Ultimate Quotable Einstein*, 414.

37. Einstein, "Paul Langevin in Memoriam" (1947), in *Out of My Later Years*, 231; *The Ultimate Quotable Einstein*, 252–253.

38. Einstein to Hans Muehsam, Mar. 30, 1954 (around a year before Einstein died), in *The Ultimate Quotable Einstein*, 454.

39. Einstein, *The Ultimate Quotable Einstein*, 454.

40. Einstein to Max Kariel, Aug. 3, 1953, in *The Ultimate Quotable Einstein*, 454.

41. Einstein to Hermann Huth, Dec. 27, 1930. Huth was vice president of the German vegetarian society the Vegetarier-Bund. In *The Ultimate Quotable Einstein*, 453.

42. Einstein, *The Philosophy of Albert Einstein*, 9.

43. Goldman, *Einstein's God*, 119.

44. Einstein, *The Philosophy of Albert Einstein*, 9.

45. Goldman, *Einstein's God*, 119–120.

46. Sayen, *Einstein in America*, 160.

47. God is even said to be soothed by the smell of burning flesh. For just a few examples, see Exodus 29:10–14, Ezekiel 39:17, and Numbers 15:24. There are countless other instances throughout the Old Testament.

48. Adultery: see Leviticus 20:10 and Deuteronomy 22:23. Worshiping other gods: see Deuteronomy 17:2–5 and also Exodus 22:20. Violating the sabbath (i.e., working on the weekend): Exodus 31:15 and again at 35:2 (note that this last injunction, to rest on the seventh day and execute anyone who wants to work, is one of the famous Ten Commandments given directly by God to Moses, supposedly a great source of ethical enlightenment for Western culture).

49. The most (in)famous passage is probably Deuteronomy 20:16–18, in which God commands the Israelites to wipe out every man, woman, and child ("leave nothing alive that breatheth") in their conquest of the cities of the Holy Land, which he has promised to them (local inhabitants be damned). And of course there's the massacre of the Egyptian army in the Red Sea in Exodus 14. This mass murder is described as "great work" (Exodus 14:31), and Moses leads the Jews in celebrating the act in song (Exodus 15). Even innocent children were fair game: prior to the egress from Egypt, Yahweh had every firstborn baby boy of the Egyptian people killed as punishment for not setting the Jews free (Exodus 13).

50. Death penalty for murder in the Old Testament: Exodus 21, Leviticus 24 and 35, Deuteronomy 19.

51. In the Gospel of John 8:1–11, for instance, Jesus intervenes in the stoning of a woman for adultery, thus saving her life, in what could be construed as a condemnation of capital punishment.

52. Matthew 5:38–40.

53. For a history of vegetarianism in the West, see Spencer, *Heretic's Feast*; for the Pythagorean diet in particular, see ch. 2.

54. Einstein to French pacifist Jacques Hadamard, Apr. 19, 1952, in *Einstein on Peace*, 565.

55. The scholar Unto Tähtinen argues that "the first reference to *ahimsa* in a moral sense may be found in *Kapisthala-Katha-Samhita*. There the non-killing of animals (*pasu-ahimsa*) is referred to in the context of sacrifice. This reference is pre-upanisadic [i.e., it is part of the earlier Vedic tradition]." See Tähtinen, *Ahimsa*, 3.

56. The Indus Valley civilization left behind some writing in carved stone inscriptions, but these are only short "sentences" or phrases, and in any case the script has yet to be deciphered. The argument for *ahimsa* in this early civilization is based instead on archaeological evidence: for example, there are carved images of a figure seated in what appears to be a meditation posture, surrounded by happy-seeming animals who appear to have no fear of this person. As I said, circumstantial at best, but the arguments are intriguing nonetheless. See Chapple, *Nonviolence*, 5–9.

57. The Chandogya Upanishad was probably written down sometime in the seventh century B.C.E., but of course its content could be much older, conveyed through oral tradition for who knows how long. T. W. Rhys Davids, an early British scholar of Buddhism and the Pali language, argued that the first real use of *ahimsa* in the way we understand it today was in Chandogya Upanishad 3.17.4. For a discussion and citation of Rhys Davids's views, see Tähtinen, *Ahimsa*, 2.

58. Chandogya Upanishad, 8.15.1. See Radhakrishnan, *The Principal Upanisads*, 512.

59. Kelsey Jo Starr, "6 Facts About Jains in India," Pew Research Center, Aug. 17, 2021, www.pewresearch.org/fact-tank/2021/08/17/6-facts-about-jains-in-india/.

60. As already mentioned, Einstein describes reading the book in a letter to Marcel Grossman, Sept. 6, 1901. Quoted in Howard, "A Peek Behind the Veil of Maya," 93.

61. "Empedocles is regularly connected in the ancient tradition with Pythagoras." Burkert, *Lore and Science*, 133.

62. Schopenhauer, "Fragments for the History of Philosophy," in *Parerga and Paralipomena*, 1:35. For full transparency, I should note that Schopenhauer's discussion here involves more than just vegetarianism. He is talking about the wider philosophy of *karma* and reincarnation (*metempsychosis* to the Greeks) and its relation to the injunction of the ethical treatment of animals. But for Schopenhauer, none of this was to be taken literally; reincarnation was only a dumbed-down version of metaphysical monism. As he wrote: "To be just, noble, and benevolent is nothing but to translate my metaphysics into actions. To say that time and space are mere forms of our knowledge, not determinations of things-in-themselves, is the same as saying that the teaching of metempsychosis, namely that 'One day you will be born again as the man whom you now injure, and will suffer the same injury,' is identical with the frequently mentioned formulae of the Brahmans, *Tat tvam asi*, 'This thou art.' All genuine virtue proceeds from the immediate and *intuitive* knowledge of the metaphysical identity of all beings.... The doctrine of metempsychosis [i.e., reincarnation]...deviates from the truth merely by transferring to the future what is already now. Thus it represents my true inner being-in-itself as existing in others only after my death, whereas the truth is that it already lives in them now." *World as Will and Representation*, 2:600–601. Schopenhauer makes it crystal clear that he thought of reincarnation as a metaphor devised by the enlightened sages to reach the

masses, who had neither the time nor the inclination to attain a subtler understanding on nonduality through spiritual practice. "For the great mass" of people, he writes, "this doctrine is too subtle; and so plain metempsychosis is preached to them as a comprehensible substitute" (2:503). It's not necessarily important, for our purposes, whether or not Schopenhauer is "right"—or even whether there can possibly be any "right" interpretation of a doctrine as diverse and complex as reincarnation. Our syllogism is simple: this is how Schopenhauer saw things; Einstein was exposed to Schopenhauer's ideas on the relation of nondualism and nonviolence early on; and since Einstein later espoused vegetarianism as a spiritual ideal, it's perfectly plausible that he originally got this idea from Schopenhauer.

63. Both quotes are from Schopenhauer's essay "On the Basis of Morality." The translation I used is at https://ivu.org/history/europe19b/schopenhauer.html.

64. Einstein, *The Ultimate Quotable Einstein*, 124.

65. From a statement made by Einstein on Feb. 11, 1948, at a memorial service in Washington, D.C., following Gandhi's assassination, in Einstein, *Einstein on Peace*, 467–468.

66. Einstein, *The Ultimate Quotable Einstein*, 124.

67. For the portrait of Gandhi, see Isaacson, *Einstein*, 438. As we've already seen throughout the book, the only other portraits that adorned his home over the years were of Schopenhauer, Maxwell, Faraday, and Newton.

68. It's quite a curious coincidence that this moniker was in fact given to Gandhi by none other than Rabindranath Tagore, the mystical nondualist poet Einstein admired so much. See "Tagore Gave Gandhi the Mahatma Title: Gujarat High Court," *India Today*, Nov. 30, 1999, www.indiatoday.in/india/story/it-was-tagore-who-gave -mahatma-title-to-gandhi-guj-hc-309623-1999-11-30.

69. Einstein to a Hindu correspondent (who had urged him to emulate Gandhi and go on a hunger strike to stop the production of the hydrogen bomb), Mar. 24, 1950, in *Einstein on Peace*, 525.

70. Einstein, *Einstein on Peace*, 525.

71. Einstein to a member of the Thoreau Society, Aug. 19, 1953, in Einstein, *Einstein on Peace*, 594.

72. Gandhi, *Autobiography*, 79. Einstein's personal copy of Gandhi's autobiography is in the Albert Einstein Archives, cat. #54C4411.

73. Gandhi belonged to the Vaishnava sect, the branch of Hinduism that focuses its worship on the god Vishnu. See Jordens, *Gandhi's Religion*, 117.

74. Quoted in Jordens, *Gandhi's Religion*, 116–117.

75. Quoted in Jordens, *Gandhi's Religion*, 73.

76. Einstein to the editor of *Kaizo* magazine in Japan (explaining his views on pacifism and nuclear weapons), Sept. 20, 1952, in *Einstein on Peace*, 584.

77. Einstein, "Gandhi's Statesmanship," in *Mahatma Gandhi: Essays and Reflections on His Life and Work*, ed. S. Radhakrishnan (George Allen & Unwin, 1939), 80.

78. Einstein, *The Ultimate Quotable Einstein*, 150.

79. Schweitzer, *Philosophy of Civilization,* 79.

80. Einstein, *The Ultimate Quotable Einstein*, 150.

81. Schweitzer, *Indian Thought and its Development*, 82–83.

82. Einstein, *The Philosophy of Albert Einstein*, 58.

83. Einstein, *The Philosophy of Albert Einstein*, 59–60.

84. Einstein, *The Philosophy of Albert Einstein*, 60.

85. Gandhi, *Wit and Wisdom*, 102.

86. Hermanns, *Einstein and the Poet*, 106.

87. Translation of verse 49 from Au-Young, *Tao Teh King*, 82, which Einstein owned. Einstein marked verse 49 in his German translation by Fiedler.

88. Einstein, "The Need for Ethical Culture," letter read to the Ethical Culture Society in New York, Jan. 1951, in *Ideas and Opinions*, 58.

7: A Conscious Freedom

1. Viereck, *Glimpses of the Great*, 452.

2. Einstein, *Ideas and Opinions*, 35.

3. Viereck, *Glimpses of the Great*, 444.

4. Einstein, *Ideas and Opinions*, 42.

5. Einstein, contribution to *Baruch Spinoza*, 29.

6. Planck, *Where Is Science Going?*, 203.

7. Spinoza to Henry Oldenburg, Nov. 20, 1665, in *Spinoza: Complete Works*, 849.

8. Einstein and Infeld, *Evolution of Physics*, 241.

9. All quotes in this paragraph from Einstein and Infeld, *Evolution of Physics*, 242.

10. Einstein and Infeld, *Evolution of Physics*, 242–243.

11. Planck, *Where Is Science Going?*, 204.

12. Planck, *Where Is Science Going?*, 203–204.

13. Planck, *Where Is Science Going?*, 203–204.

14. Hermanns, *Einstein and the Poet*, 98.

15. Einstein, *The Philosophy of Albert Einstein*, 55.

16. Einstein, *The Philosophy of Albert Einstein*, 11.

17. For a scintillating survey of earlier thinkers' thoughts on the unconscious, complete with many direct quotations, see Whyte, *The Unconscious Before Freud*.

18. Freud denied any direct influence by Schopenhauer or Nietzsche, but historians know these claims can't be taken seriously. There are countless examples noted in the scholarly literature, but for just one recent study, see Cybulska, "Freud's Burden of Debt to Nietzsche and Schopenhauer."

19. Schopenhauer, *World as Will and Representation*, 1:113–114.

20. Viereck, *Glimpses of the Great*, 441.

21. Viereck, *Glimpses of the Great*, 442.

22. Viereck, *Glimpses of the Great*, 442.

23. Viereck, *Glimpses of the Great*, 441.

24. Viereck, *Glimpses of the Great*, 441.

25. Allegedly the words of Einstein, but no source is given. Quoted in Goldman, *Einstein's God*, 60.

26. Einstein to Maxim Gorki, Sept. 29, 1932, in *Einstein on Peace*, 204.

27. Goldman, *Einstein's God*, 55.

28. Allegedly from one of Einstein's letters, but no source, recipient, or date is provided. Quoted in Goldman, *Einstein's God*, 55.

29. Quoted in Goldman, *Einstein's God*, 57.

30. Viereck, *Glimpses of the Great*, 452.

31. Einstein, "Why Socialism?" (1949), in *Out of My Later Years*, 126.

32. Einstein, *Out of My Later Years*, 126.

33. Einstein, *Ideas and Opinions*, 35.

34. The first quotation is in Einstein, *Einstein on Peace*, 506. The second is from an address Einstein gave at a mass meeting of some ten thousand people gathered to help scholarly refugees escape from the rising Nazi state (the event was held in London on Oct. 3, 1933, at the Royal Albert Hall); *Einstein on Peace*, 239.

35. Einstein, *Ideas and Opinions*, 35.

36. Einstein to a nineteen-year-old student at Rutgers University, Dec. 3, 1950, in Dukas and Hoffmann, *Albert Einstein: The Human Side*, 27.

37. Einstein, *Ideas and Opinions*, 35.

38. Quoted in Dukas and Hoffmann, *Albert Einstein: The Human Side*, 27.

39. Einstein, "The Goal of Human Existence" (1943), in *Out of My Later Years*, 260.

40. Einstein, *The Ultimate Quotable Einstein*, 334.

41. Einstein, "Science and God," 379.

42. Goldman, *Einstein's God*, 58.

43. Goldman, *Einstein's God*, 59–60.

44. Goldman, *Einstein's God*, 57–58.

45. David Fideler, in his introduction to Guthrie and Fideler, *Pythagorean Sourcebook*, 37.

46. Einstein, "On Academic Freedom: The Gumbel Case," Apr. 28, 1931, in *Einstein on Politics*, 464.

47. Einstein, *Out of My Later Years*, 127.

48. Quoted in Goldman, *Einstein's God*, 117.

49. The opening paragraph from Einstein, "What I Believe."

50. Einstein, *The Ultimate Quotable Einstein*, 482. The quotation is unsourced and placed by Calaprice under the heading "Probably Not by Einstein" in the final "Attributed to Einstein" part of her book. As with some other quotes that are similarly dubious, even if Einstein did not say these exact words, in this case they certainly capture the essence of his beliefs.

51. Einstein, *The Ultimate Quotable Einstein*, 341.

52. Einstein, *The Ultimate Quotable Einstein*, 328.

53. Hermanns, *Einstein and the Poet*, 72.

54. Hermanns, *Einstein and the Poet*, 72.

55. Einstein, "Human Rights," message delivered to the Decalogue Society, Feb. 20, 1954, in *Einstein on Peace*, 600.

56. Einstein to a correspondent in California, Nov. 9, 1953, in *Einstein on Peace*, 596.

57. Einstein, "The Need for Ethical Culture," Jan. 1951, in *Ideas and Opinions*, 57.

58. Einstein, *Ideas and Opinions*, 57.

59. From Bertrand Russell's Peace Manifesto, which Einstein signed shortly before he died, in *The Ultimate Quotable Einstein*, 285.

60. Hermanns, *Einstein and the Poet*, 89.

61. Einstein to Hedwig Born, Aug. 31, 1919, in *The Ultimate Quotable Einstein*, 418.

62. Hermanns, *Einstein and the Poet*, 104.

63. Hermanns, *Einstein and the Poet*, 66.

64. Hermanns, *Einstein and the Poet*, 66.

65. Einstein, *The Philosophy of Albert Einstein*, 59.

66. Einstein, *The Philosophy of Albert Einstein*, 59.

67. Hermanns, *Einstein and the Poet*, 68.

68. Einstein's main arguments against traditional religion are discussed in detail in Chapters 1 and 2.

69. Einstein, *The Philosophy of Albert Einstein*, 59.

70. Einstein, *The Philosophy of Albert Einstein*, 59.

71. *New York Times*, Apr. 25, 1929, 60, in Jammer, *Einstein and Religion*, 48.

72. The dinner in London where Einstein met the archbishop and the conversation they had are described in Frank, *Einstein*, 189–190.

73. For an excellent article tracing the history of these claims, and their lack of credibility, see David Greenberg, "It Didn't Start with Einstein," *Slate*, February 3, 2000, https://slate.com/news-and-politics/2000/02/it-didn-t-start-with-einstein.html.

74. Walter Isaacson, "Who Mattered and Why," *Time*, Dec. 31, 1999, 60.

75. Einstein, *The Philosophy of Albert Einstein*, 58.

76. Einstein, *The Philosophy of Albert Einstein*, 58.

77. Einstein, "The Laws of Science and the Laws of Ethics," in *Out of My Later Years*, 114–115.

78. Einstein, "Science and God," 374.

79. Einstein to a member of a California law firm, May 16, 1951, in Einstein, *Einstein on Peace*, 556.

80. Einstein, "Science and God," 374.

81. Einstein, "Science and God," 375.

82. Hermanns, *Einstein and the Poet*, 66.

83. From an entry Einstein made in a neighbor's album near his summer home in Caputh in 1932. Quoted in Dukas and Hoffmann, *Albert Einstein: The Human Side*, 30.

84. Hermanns, *Einstein and the Poet*, 57.

8: Part of Infinity

1. Einstein, *The Ultimate Quotable Einstein*, 482.

2. Einstein to Robert Marcus, Feb. 12, 1950, in *The Ultimate Quotable Einstein*, 339–340 (Albert Einstein Archives doc. 60-424).

3. Quoted in Goldman, *Einstein's God*, 89.

4. Einstein, "Why Socialism?" (1949), in *Out of My Later Years*, 128.

5. Hermanns, *Einstein and the Poet*, 109.

6. Einstein, "Science and God," 379.

7. Hermanns, *Einstein and the Poet*, 27 (said during a discussion of Spinoza's idea of the divine).

8. Spinoza, *Ethics*, book V, prop. 31, scholium, in *Spinoza: Complete Works*, 377.

9. Schopenhauer, *The World as Will and Representation*, 1:178–179.

10. Schopenhauer, *The World as Will and Representation*, 1:411.

11. Einstein, *The Philosophy of Albert Einstein*, 9.

12. Einstein, *The Ultimate Quotable Einstein*, 113.

13. Einstein to the queen of Belgium, Jan. 9, 1939, in *Einstein on Peace*, 282.

14. Dukas and Hoffmann, *Albert Einstein: The Human Side*, 23.

15. Einstein to the queen of Belgium, Jan. 12, 1953, quoted in Hoffmann and Dukas, *Albert Einstein: Creator and Rebel*, 261, and in *Einstein on Peace*, 591.

16. Dukas and Hoffmann, *Albert Einstein: The Human Side*, 38.

17. Dukas and Hoffmann, *Albert Einstein: The Human Side*, 38.

18. Einstein, contribution to *Baruch Spinoza*, 28–29.

19. Viereck, *Glimpses of the Great*, 448. Note that these words of Einstein's are a complete contradiction of Max Jammer's obtuse insistence that "the only connecting link between Spinoza's philosophy and Einstein's...[is] determinism" (*Einstein and Religion*, 45), already mentioned in Chapter 1.

20. All quotes here are from the Chandogya Upanishad, as translated in Browne, *The World's Great Scriptures*, 75.

21. Lao Tzu, *Tao Te Ching*, 84.

22. Photius, *Life of Pythagoras*, §15, in Guthrie and Fideler, *Pythagorean Sourcebook*, 139.

23. Schopenhauer, *The World as Will and Representation*, 1:205.

24. Einstein to Max Born and Hedwig Born, Apr. 12, 1949, in Born, *The Born-Einstein Letters*, 178. See also the final section of Chapter 6.

25. Einstein, "Good and Evil," in *The World as I See It*, 18; also in *The Philosophy of Albert Einstein*, 33.

26. Quoted in Goldman, *Einstein's God*, 119.

27. Quoted in Goldman, *Einstein's God*, 119.

28. Einstein, *Ideas and Opinions*, 48.

29. Einstein, *The Ultimate Quotable Einstein*, 503.

30. Hoffmann and Dukas, *Albert Einstein: Creator and Rebel*, 94.

31. Quoted in Pais, *Einstein Lived Here*, 102.

32. Hedi Born to Einstein, Oct. 9, 1944, in Born, *The Born-Einstein Letters*, 150.

33. Einstein, *The Ultimate Quotable Einstein*, 57.

34. Einstein, *The Travel Diaries of Albert Einstein*, 258.

35. Einstein, *The Travel Diaries of Albert Einstein*, 258.

36. Einstein to the queen of Belgium, Aug. 12, 1939, in *Einstein on Peace*, 285.

37. Hermanns, *Einstein and the Poet*, 61, 117.

38. In the Bhagavad Gita (III, 3–9), for instance, the god Krishna tells Arjuna of "the path of knowledge (*jñāna-yoga*) for men of discrimination." See Deutsch and Dalvi, *The Essential Vedanta*, 65. Similarly, one of the early Upanishads, the Amritabindu Upanishad, praises the path of *jñāna*: "Knowing 'I am the Absolute,' the Absolute is surely attained [Verse 8]... The sage who, after studying the books, is intent on that [Absolute] through wisdom (*jñāna*) and knowledge (*vijñāna*) should discard all books, even as the husk [is discarded by a person] seeking the grain [Verse 18]." As quoted in Feuerstein, *The Yoga Tradition*, 35–36.

39. In the words of Nalini Kanta Brahma in his *Philosophy of Hindu Sadhana* (1932), quoted in Feuerstein, *The Yoga Tradition*, 32.

40. Shankara, *Crest-Jewel*, 72.

41. Shankara, *Crest-Jewel*, 70.

42. Shankara, *Crest-Jewel*, 82.

43. Shankara, *Crest-Jewel*, 36.

44. Shankara, *Crest-Jewel*, 36.

45. Shankara, *Crest-Jewel*, 33.

46. Shankara, *Crest-Jewel*, 56.

47. Shankara, *Crest-Jewel*, 34–35.

48. Shankara, *Crest-Jewel*, 43.

49. For instance, see Schopenhauer, *The World as Will and Representation*, 2:508, 607. Notably, Schopenhauer directly refers to Shankara in connection with "the Vedantists" (2:508) and "the Vedanta philosophy" (2:607), so there can be little doubt that Einstein was aware of this school of thought and that Shankara was one of its major exponents.

50. Schopenhauer, *The World as Will and Representation*, 2:508.

51. Schopenhauer, *The World as Will and Representation*, 2:475.

52. Spinoza, *Ethics*, part II, proposition 40, scholium 2, in *Spinoza: Complete Works*, 267.

53. Spinoza, *Ethics*, part V, proposition 25, proof, in *Spinoza: Complete Works*, 375.

54. Spinoza, *Ethics*, part V, proposition 32, corollary, in *Spinoza: Complete Works*, 377.

55. Spinoza, *Ethics*, part V, proposition 20, scholium, in *Spinoza: Complete Works*, 373.

56. Spinoza, *Ethics*, part V, proposition 14, in *Spinoza: Complete Works*, 371.

57. Spinoza, *Ethics*, part II, proposition 44, corollary 2, proof, in *Spinoza: Complete Works*, 270.

58. Spinoza, *Ethics*, part II, proposition 44, corollary 2, in *Spinoza: Complete Works*, 270.

59. Spinoza, *Treatise on the Emendation of the Intellect*, para. 11, in *Spinoza: Complete Works*, 5.

60. Spinoza, *Ethics*, part V, proposition 20, scholium, in *Spinoza: Complete Works*, 373; Spinoza, *Treatise on the Emendation of the Intellect*, para. 11, in *Spinoza: Complete Works*, 5–6.

61. Spinoza, *Ethics*, part V, proposition 36, in *Spinoza: Complete Works*, 378.

62. Einstein, "Science and God," 375.

63. Einstein, *Ideas and Opinions*, 56–57.

64. Einstein, *Ideas and Opinions*, 52–53.

65. From a little epigram Einstein inscribed on an etching of himself, written around 1920. Quoted in Dukas and Hoffmann, *Albert Einstein: The Human Side*, 24.

66. Einstein, *Ideas and Opinions*, 52.

67. Einstein, *Ideas and Opinions*, 53.

68. In the last major interview Einstein ever gave, his interviewer brought up Newton's shameful behavior toward his contemporaries Robert Hooke and Gottfried

Wilhelm Leibniz, to whom he would not concede the slightest bit of credit for either pre-figuring or co-discovering any of his ideas. Einstein agreed that Newton's attitude was abhorrent. "That, alas, is vanity," he said. "You find it in so many scientists. You know, it has always hurt me to think that Galilei did not acknowledge the work of Kepler." I. B. Cohen, "Einstein's Last Interview," *Scientific American* 193, no. 1 (1955): 68–73, esp. 69. The key quote on the vanity of Newton and Galileo is also in *The Ultimate Quotable Einstein*, 122.

69. James Clerk Maxwell, quoted in Stone, *Einstein and the Quantum*, 31.

70. Maxwell, quoted in Stone, *Einstein and the Quantum*, 32.

71. Gutfreund and Renn, *Einstein on Einstein*, 158.

72. Schopenhauer, *The World as Will and Representation*, 1:352.

73. Gutfreund and Renn, *Einstein on Einstein*, 158.

74. Einstein, *Ideas and Opinions*, 52.

75. Hermanns, *Einstein and the Poet*, 117.

76. Einstein, *Ideas and Opinions*, 52–53.

77. To be fully transparent: in Faraday's time it wasn't yet understood that the electromagnetic force was what held the body's organic compounds together. But we now know that the extremely powerful attractive force between negatively charged electrons and positively charged atomic nuclei (so-called chemical bonding) is what keeps our body's molecules tough and stable over time.

78. Einstein, "Maxwell's Influence on the Evolution of the Idea of Physical Reality" (1931), in *Ideas and Opinions*, 294.

79. Einstein, *Ideas and Opinions*, 294.

80. Einstein, *Ideas and Opinions*, 294.

81. Einstein to Max Born, Nov. 9, 1919, in *The Ultimate Quotable Einstein*, 6.

82. Interview with Einstein, *London Daily Chronicle*, Jan. 26, 1929, in Clark, *Einstein*, 407.

83. Quoted in Clark, *Einstein*, 407.

84. Einstein, interviewed by Alfred Stern, *Contemporary Jewish Record* 8 (1945): 245–249, in *The Ultimate Quotable Einstein*, 395.

85. Gutfreund and Renn, *Einstein on Einstein*, 166.

86. A. Einstein and J. Grommer, "Beweis der Nichtexistenz eines überall regulären zentrisch symmetrischen Feldes nach der Feldtheorie von Kaluza," *Scripta* (Jerusalem University) 1, no. 7 (1923): 1–5.

87. Einstein, "Einheitliche Feldtheorie von Gravitation und Elektrizität," *Sitzungsberichte der Preussischen Akademie der Wissenschaften*, 1925, 414–419, quote from the English translation by A. Unzicker and T. Case, 1, at http://u2.lege.net/cetinbal/PDFdosya/Gravitation_Electricity.pdf.

88. Einstein's paper was published in July 1925. The quotes are from two letters Einstein later wrote to his friend Paul Ehrenfest, dated Aug. 18 and Sept. 20, 1925, respectively. Quoted in Pais, *Subtle Is the Lord*, 344.

89. Einstein to Heinrich Zangger, Feb. 27, 1938, in Fölsing, *Albert Einstein*, 552.

90. I won't waste space listing them all here, but Einstein's obsession spanned some four decades. Einstein was working on unified field theories at least as early as 1917; his first paper on unified field theory appeared in 1923, and his very last scientific

paper, published posthumously in 1955, also dealt with this topic. For a comprehensive chronology and semi-technical treatment of these many papers, see Pais, *Subtle Is the Lord*, ch. 17, esp. 341–351. A complete list of Einstein's nearly three hundred scientific publications can be found at https://en.wikipedia.org/wiki/List_of_scientific _publications_by_Albert_Einstein.

91. Born, "Einstein's Statistical Theories," 163.

92. Hermanns, *Einstein and the Poet*, 140.

93. Quoted in Miller, *Deciphering the Cosmic Number*, 24.

94. Pauli's first paper, written when he was only eighteen years old, was "Über die Energiekomponenten des Gravitationsfeldes," *Physikalische Zeitschrift* 20, no. 25 (1919).

95. Einstein was twenty-six by the time his papers on special relativity and the photoelectric effect were published in late 1905, but most of the creative work had been done earlier in the year, prior to his twenty-sixth birthday.

96. Pauli's exclusion principle provides the primary explanation for the structure and stability of matter at the atomic scale: why atoms, made up of negative and positive particles that desperately want to get together, don't just collapse in on themselves, but instead stay stable for billions of years. See Enz, *No Time to Be Brief*, 119–129. Pauli also played a key role in understanding spin, an irreducible property of elementary particles that's just as fundamental as electric charge or mass; see Enz, *No Time to Be Brief*, 106–119. And with his incredible intuition and audacity, Pauli was even able to predict the existence of a new elementary particle, the neutrino, decades before it was discovered in experiments; see Enz, *No Time to Be Brief*, 209–239.

97. Unfortunately, the speech seems to have been impromptu rather than prepared, and beyond Pauli's vague reminiscences no record of it survives (as far as I'm aware). The staff at the Albert Einstein Archives confirmed for me that there is no draft of the speech in Einstein's personal papers, and even his lifelong assistant, Helen Dukas, didn't think any written record existed (Chaya Becker, personal communication, Feb. 26, 2023). Pauli also believed the speech was lost to history. "Unfortunately, there are no records of Einstein's speech (it was improvised and a manuscript does not exist either)," he wrote in a letter to Max Born dated Apr. 24, 1955, the week after Einstein died. The letter is quoted, in German, in Pauli, *Writings on Physics and Philosophy*, 114 (the English above is my own translation). The quote about Einstein being "at the end of his wisdom" is Alice Calaprice's commentary on the content of Einstein's speech, in *The Ultimate Quotable Einstein*, 523. As source, Calaprice quotes Armin Hermann, "Einstein und die Österreicher," *Plus Lucis*, February 1995, 20–21. The quote about "the great theory" is Hermann's paraphrase of Einstein's words in the speech (20).

98. Letter from Pauli to their mutual friend Max Born, Apr. 25, 1955 (one week after Einstein died). This English translation is provided in Karl von Meyenn and Engelbert Schucking, "Wolfgang Pauli," *Physics Today* 54, no. 2 (2001): 43. It's fair to assume the translation is accurate, given that von Meyenn is the editor of Pauli's complete correspondence in German, published in six volumes between 1979 and 2000 (*Wolfgang Pauli, Wissenschaftlicher Briefwechsel* [Springer-Verlag]). It's important to note that this feeling of being Einstein's appointed heir was not mere ego or vanity on Pauli's part; his peers, perhaps a little grudgingly, also acknowledged his right to the throne. Max Born, for instance (also a Nobel laureate in physics, and a close friend of both Einstein and

Pauli), wrote that Pauli "became a close friend of Einstein's and regarded himself, probably with some justification, as the designated 'successor' in theoretical physics." See Born, *The Born-Einstein Letters*, 212.

99. Pauli, "Albert Einstein and the Development of Physics" (1958), in *Writings on Physics and Philosophy*, 123.

100. Quoted in Miller, *Deciphering the Cosmic Number*, 261; Pauli to Heisenberg, Jan. 1958, in Heisenberg, *Physics and Beyond*, 234.

101. Quoted in Miller, *Deciphering the Cosmic Number*, 260.

102. Niels Bohr, J. Robert Oppenheimer, Freeman Dyson, and Richard Feynman all apparently attended lectures on the as yet unpublished theory. Miller, *Deciphering the Cosmic Number*, 261–263.

103. The origin of these epithets is discussed in Miller, *Deciphering the Cosmic Number*, 107–108.

104. Quoted in Miller, *Deciphering the Cosmic Number*, 262.

105. According to Heisenberg's recollections; Heisenberg, *Physics and Beyond*, 235.

106. Quoted in Miller, *Deciphering the Cosmic Number*, 263.

9: A Magnificent Synthesis

1. From a draft of a message Einstein wrote on his fiftieth birthday, Mar. 14, 1929, in *Einstein on Peace*, 95–96.

2. Einstein, "Science and God," 375.

3. Einstein, address at Columbia University, quoted in Henry Margenau, "Einstein's Conception of Reality," in *Albert Einstein: Philosopher-Scientist*, ed. P. A. Schilpp, 249. Tudor, 1951.

4. Einstein, "Science and God," 375.

5. Einstein, "Science and God," 375.

6. Einstein, "Physics and Reality" (1936), in *Ideas and Opinions*, 355.

7. Einstein, "Science and God," 378.

8. Einstein, "On the Method of Theoretical Physics" (1933), in *Ideas and Opinions*, 297; Dukas and Hoffmann, *Albert Einstein: The Human Side*, 38.

9. Quoted in Runes, *Spinoza Dictionary*, v–vi.

10. Einstein, contribution to *Baruch Spinoza*, 28–29.

11. From Einstein's preface to Planck, *Where Is Science Going?*, 11–12.

12. Quoted in Planck, *Where Is Science Going?*, 12.

13. Einstein to his personal physician Janos Plesch, in Fölsing, *Albert Einstein*, 601.

14. Einstein, *Ideas and Opinions*, 51.

15. Planck, *Where Is Science Going?*, 11.

16. Dukas and Hoffmann, *Albert Einstein: The Human Side*, 38.

17. Planck, *Where Is Science Going?*, 11–12.

18. Planck, *Where Is Science Going?*, 10–11.

19. Max Born and J. Robert Oppenheimer, "Zur Quantentheorie der Molekeln," *Annalen der Physik* 389, no. 20 (1927): 457–484.

20. Schrödinger, *What Is Life?*

21. In his memoirs, James Watson recalls how he had "become focused on the gene through a reading that winter of Erwin Schrödinger's thin book *What Is Life? . . .*

Schrödinger elegantly laid out how genes were the most important feature of life, since they maintained its continuity by carrying hereditary information from one generation to the next....Schrödinger's exaltation of the gene would lead me to a life of studying genetics." See Watson, *Avoid Boring People*, 28. Francis Crick likewise gave Schrödinger's book part of the credit for getting him interested in pursuing a career in physical biology. See Crick, *What Mad Pursuit*, 18.

22. Hermanns, *Einstein and the Poet*, 61.

23. From a draft of a reply to the Adlerian German psychotherapist H. Freund, who'd invited Einstein to participate in a study involving the psychoanalysis of important people. It's not clear whether the reply was ever actually sent. Einstein, *The Ultimate Quotable Einstein*, 448 (Albert Einstein Archives, doc. 46-304).

24. Viereck, "What Life Means to Einstein," *Saturday Evening Post*, Oct. 26, 1929, 117, 114. The interview is reprinted, with some minor differences, in Viereck, *Glimpses of the Great*.

25. Einstein to Freud, Mar. 22, 1929, in Fölsing, *Albert Einstein*, 609.

26. Hermanns, *Einstein and the Poet*, 103.

27. Pauli to Jung, Feb. 27, 1953, in Meier, *Atom and Archetype*, 90.

28. Pauli to Jung, May 17, 1952, in Meier, *Atom and Archetype*, 81–82.

29. Heisenberg, *Across the Frontiers*, 30.

30. Pauli had many dreams that illustrated this ambivalence, and in one letter to Jung he analyzed his own resistance in detail. "First of all," he wrote, referring to a dream he'd included with his letter, "there is the *motif of the auditorium*, with *strangers*, in front of whom I am to hold *lectures*. This has cropped up in previous dreams and is closely linked to dreams that I had been offered *a new professorship* but had not yet accepted....The motif of the not-yet-accepted professorship seems to me very important, for it shows the resistance of the conscious to the 'professorship.' The unconscious is rebuking me for having kept something specific from the public, something akin to a confession that I had not accepted my appointment out of conventional forms of resistance." Pauli to Jung, Feb. 27, 1953, in Meier, *Atom and Archetype*, 89. Many similar instances of Pauli's resistance to becoming a public figure speaking on behalf of the wisdom of the unconscious are documented in *Atom and Archetype*.

31. For the origin of this epithet, see Miller, *Deciphering the Cosmic Number*, 108.

32. This dark period in Pauli's life, and how his personal crisis led him to Jung, is discussed in detail in Miller, *Deciphering the Cosmic Number*, 109–117, and in Enz, *No Time to Be Brief*, 240–249. See also Bair, *Jung*, 365–366.

33. See Enz, *No Time to Be Brief*, 244. For much more on Jung's Bollingen Tower, see Ziolkowski, *View from the Tower*.

34. Jung was fluent in both ancient Greek and Latin and so was able to read these texts in the original. He provides a detailed description of his method of actively entering into the unconscious in his autobiography. "In order to seize hold of the fantasies, I frequently imagined a steep descent. I even made several attempts to get to the very bottom. The first time I reached, as it were, a depth of about a thousand feet; the next time I found myself at the edge of a cosmic abyss. It was like a voyage to the moon, or a descent into empty space....I had the feeling that I was in the land of the dead."

Jung, *Memories, Dreams, Reflections*, 181. Several scholars have pointed out that Jung's method shares striking similarities with the ancient Greek method of *enkoimesis* (incubation) used to effect a *katabasis* (going down) into the underworld in order to obtain wisdom and healing from the gods. As the classicist Peter Kingsley points out: "The underworld for Jung is the same as the underworld for Parmenides... on the most basic, instinctive level scholars have had such a hard time accepting that someone like Parmenides—founder of western logic—or Pythagoras would have been forced to visit the world of the dead while still alive.... And this is why until recently Jung's assistants, editors, publishers worked so hard to cut any mention of the underworld out of his writings: is why they tried to suppress the direct parallel he himself was so keen to insist on, between his own otherworldly journey and the traditional Greek descents into the world of the dead." See Kingsley, *Catafalque*, 62–63. This parallel was also developed in detail in a book by one of Jung's closest disciples: see Meier, *Ancient Incubation and Modern Psychotherapy*. And for an intriguing collection of Jung's own writings on the origins and methods of active imagination, see Jung, *Jung on Active Imagination*.

35. Quoted in Bair, *Jung*, 366.

36. Jung, "Individual Dream Symbolism in Relation to Alchemy" (1936), in *Psychology and Alchemy*, 42.

37. Jung, *Psychology and Alchemy*, 42.

38. Jung, *Psychology and Alchemy*, 43.

39. Jung, quoted in Enz, *No Time to Be Brief*, 243; Jung, *Psychology and Alchemy*, 42; Zabriskie, "Jung and Pauli," xxxiii.

40. Jung, *Psychology and Alchemy*, 42.

41. Jung, "The Tavistock Lectures: On the Theory and Practice of Analytical Psychology" (1968), in Meier, *Atom and Archetype*, xxxiii.

42. For details of how this came about, see Meier, *Atom and Archetype*, xxxii–xxxiii. Jung originally presented the work as a series of lectures given in 1936 ("Individual Dream Symbolism in Relation to Alchemy"). The study is reprinted in full in Jung, *Psychology and Alchemy*, 39–223.

43. These regular in-person meetings began in 1932. See Bair, *Jung*, 366.

44. Jung sent Pauli a draft of his book on synchronicity and gratefully accepted Pauli's comments and criticisms. "Many thanks for your kind letter and for the time and trouble you have taken with my manuscript," Jung wrote. "Your opinions are very important to me." Jung to Pauli, Nov. 30, 1950, in Meier, *Atom and Archetype*, 59. See also letters 43J and 45P.

45. *Natureklärung und Pysche*, originally published in German in 1952. It was translated into English in 1955 as *The Interpretation of Nature and the Psyche*.

46. "I truly welcome your wish to found an institute with the objective of cultivating and promoting the field of research that you have inaugurated; and I give my consent to my name being on the list of sponsors." Pauli to Jung, Dec. 23, 1947, in Meier, *Atom and Archetype*, 32.

47. All quotations of Jung's words here are from Zabriskie, "Jung and Pauli," xxx.

48. The two preceding quotations of Jung's words are again from Zabriskie, "Jung and Pauli," xxxii.

49. In contrast to Jung's glowing assessment of Einstein, Einstein was at first unimpressed by the "vague, imprecise notions" of Jung's analytical psychology, which he considered "worthless." He felt that "if there has to be a psychiatrist, I should prefer Freud. I do not believe in him, but I love very much his concise style and his original, although rather extravagant, mind." From an entry in Einstein's travel diary, Dec. 6, 1931, when he was en route to the United States, in Einstein, *Einstein on Peace*, 185; also in *The Ultimate Quotable Einstein*, 121. But later in life, Einstein grudgingly agreed that "Jung is a great man" and admitted to being intrigued by Jung's notion of synchronicity. "I agree with him," Einstein said, "that causal methods in physics are not enough to explain the laws of the universe." See Hermanns, *Einstein and the Poet*, 103–104.

50. Pauli, "Unpublished Essay," 180; Pauli, "Science and Western Thought" (1955), in *Writings on Physics and Philosophy*, 145.

51. Jung, *Memories, Dreams, Reflections*, 208.

52. Pauli, "Science and Western Thought," 148.

53. Jung, "Archetypes of the Collective Unconscious" (1954), in *The Archetypes and the Collective Unconscious*, 36–37.

54. Jung, *Memories, Dreams, Reflections*, 182–184.

55. Jung, *The Archetypes and the Collective Unconscious*, 37.

56. Pauli to Jung, May 24, 1937, in Meier, *Atom and Archetype*, 19.

57. Discussed in detail in Chapter 4.

58. Pauli, "Unpublished Essay," 186.

59. Pauli, "Unpublished Essay," 183, 186.

60. Quoted in Miller, *Deciphering the Cosmic Number*, xv.

61. There are countless examples in the dreams Pauli shared with Carl Jung. See Meier, *Atom and Archetype*.

62. Pauli, "Unpublished Essay," 182.

63. Pauli, "Unpublished Essay," 186.

64. This dream is described in van Erkelens, "Wolfgang Pauli's Dialogue with the Spirit of Matter," 37.

65. From a letter of Pauli's dated Apr. 24, 1955, reproduced in German in *Writings on Physics and Philosophy*, 114. My translation.

66. From a dream dated May 20, 1955, around one month after Einstein's death. The dream report and Pauli's commentary on it are reproduced in Meier, *Atom and Archetype*, 148–149.

67. Pauli, "Unpublished Essay," 180; Pauli to Jung, Oct. 30, 1938, in Meier, *Atom and Archetype*, 22.

68. Pauli to Jung, April 28, 1934, in Meier, *Atom and Archetype*, 26.

69. Jung, "Individual Dream Symbolism in Relation to Alchemy" (1936), in *Psychology and Alchemy*, 49.

70. Jung, *Psychology and Alchemy*, 50.

71. Jung, *Psychology and Alchemy*, 50.

72. Pauli to Jung, May 24, 1934, in Meier, *Atom and Archetype*, 27.

73. Meier, *Atom and Archetype*, 27.

74. Meier, *Atom and Archetype*, 27.

75. Meier, *Atom and Archetype*, 27.

76. Meier, *Atom and Archetype*, 27.

77. Meier, *Atom and Archetype*, 27.

78. Pauli, "Two Lectures by Pauli at the Psychological Club of Zurich" (1948), in Meier, *Atom and Archetype*, 208.

79. Pauli to Jung, May 17, 1952, in Meier, *Atom and Archetype*, 81–82.

80. Pauli to Jung, Oct. 23, 1956, in Meier, *Atom and Archetype*, 141.

81. Pauli to Jung, Feb. 27, 1953, in Meier, *Atom and Archetype*, 87.

82. Pauli, "Science and Western Thought," 139.

83. Pauli, "Science and Western Thought," 139.

84. Pauli, "Science and Western Thought," 139–140.

85. Pauli, "Science and Western Thought," 140.

86. Pauli, "Ideas of the Unconscious from the Standpoint of Natural Science and Epistemology" (1954), in *Writings on Physics and Philosophy*, 160; Pauli to Jung, Mar. 31, 1953, in Meier, *Atom and Archetype*, 106.

87. Pauli to Jung, Mar. 31, 1953, in Meier, *Atom and Archetype*, 109; Pauli to Jung, May 27, 1953, in Meier, *Atom and Archetype*, 119.

88. Meier, *Atom and Archetype*, 119.

89. Pauli to Jung, Feb. 27, 1953, in Meier, *Atom and Archetype*, 90; Pauli, "Science and Western Thought," 146.

90. Jung to Pauli, Oct. 24, 1953, in Meier, *Atom and Archetype*, 125.

91. Pauli, "Science and Western Thought," 145.

92. Pauli, "Matter" (1954), in *Writings on Physics and Philosophy*, 34.

93. Hermanns, *Einstein and the Poet*, 70.

94. Pauli, "Theory and Experiment" (1952), in *Writings on Physics and Philosophy*, 125.

95. The heart of Plato's discussion of *anamnesis* is found in *Meno*, §81–86. My quotations are from the translation by G. M. A. Grube, in *Plato: Complete Works*, 870–897. All quotations are from 880–886.

96. For a book-length treatment of Plato's enormous debt to the Pythagoreans, see Kingsley, *Ancient Philosophy, Mystery, and Magic*.

97. Cameron, *The Pythagorean Background of the Theory of Recollection*, 95.

98. Pauli, "Science and Western Thought," 145.

99. Pauli, "Unpublished Essay," 180.

100. Pauli, "Unpublished Essay," 180.

101. Quoted in van Erkelens, "Wolfgang Pauli's Dialogue with the Spirit of Matter," 34–36.

102. Pauli to Jung, May 27, 1953, in Meier, *Atom and Archetype*, 122.

103. Pauli to Jung, Dec. 23, 1947, in Meier, *Atom and Archetype*, 32; Pauli to Jung, Oct. 23, 1956, in Meier, *Atom and Archetype*, 142.

104. Pauli, "Science and Western Thought," 147.

105. The poignant story of Pauli's final days is told in Enz, *No Time to Be Brief*, 533–540.

106. According to Pauli's wife, in a private communication to Carl Jung's personal secretary, Aniela Jaffé, quoted in van Erkelens, "Wolfgang Pauli's Dialogue with the Spirit of Matter," 52.

10: The Eternal Enigma

1. Viereck, *Glimpses of the Great*, 448.

2. Einstein, *The Ultimate Quotable Einstein*, 344.

3. Einstein, *The Philosophy of Albert Einstein*, 43.

4. Einstein to Max von Laue, Feb. 3, 1955. Einstein died two months later, on April 18. Quoted in Fölsing, *Albert Einstein*, 740.

5. Einstein apparently said this to his friend Otto Stern. Quoted in Pais, *Subtle Is the Lord*, 9.

6. Einstein, "Physics and Reality" (1936), in *Ideas and Opinions*, 352.

7. For detailed accounts of Einstein's various objections over the years, see Pais, *Subtle Is the Lord*, 440–446, and Stone, *Einstein and the Quantum*, 268–278.

8. The major shift in our model of the atom was inaugurated by Werner Heisenberg's "matrix mechanics" paper: "Über quantentheoretische Umdeutung kinematischer und mechanischer Beziehungen," *Zeitschrift für Physik* 33, no. 1 (1925): 879–893. Only a few months later, Erwin Schrödinger published an alternative model, in which he explained electrons as standing waves oscillating within the atom's interior. And then along came Max Born, who argued that Schrödinger's wave function shouldn't be seen as an actual physical entity within the atom. Rather, the amplitude of the wave function only indicated the *probability* of finding an electron at any particular point (more specifically, it's the square of the wave's amplitude that indicates the probability). With this probabilistic interpretation of the electron's existence, the crucifixion of the old concept of the atom was complete. See Born, "Zur Quantenmechanik der Stoßvorgänge," *Zeitschrift für Physik* 37, no. 12 (1926): 863–867.

9. Quoted in Heisenberg, *Physics and Beyond*, 41.

10. Einstein, "Maxwell's Influence on the Evolution of the Idea of Physical Reality" (1931), in *Ideas and Opinions*, 291, 295.

11. Einstein, "Remarks to the Essays Appearing in This Collective Volume," 667.

12. This conversation with Einstein was reconstructed by Heisenberg from memory several decades after the fact, so all the following quotations should be taken with a grain of salt. The full account is given in Heisenberg, *Physics and Beyond*, 58–69.

13. Einstein, *Ideas and Opinions*, 300.

14. Quoted in Heisenberg, *Physics and Beyond*, 66.

15. Quoted in Heisenberg, *Physics and Beyond*, 65.

16. Quoted in Heisenberg, *Physics and Beyond*, 66.

17. Stone, *Einstein and the Quantum*, 3.

18. Einstein, Podolsky, and Rosen, "Can Quantum-Mechanical Description of Physical Reality Be Considered Complete?," 780.

19. Goldman, *Einstein's God*, 66.

20. So says Fölsing, *Albert Einstein*, 693.

21. Planck first presented this idea in a lecture to the German Physical Society on Dec. 14, 1900. As science historian Helge Kragh has pointed out, however, "the importance ascribed to his work is largely a historical reconstruction." At the time, no one (including Planck) really realized the revolutionary implications of what he'd done. For a fascinating overview of Max Planck's role in the birth of quantum physics, see Kragh, "Max Planck: The Reluctant Revolutionary."

22. Einstein, "Über einen die Erzeugung und Verwandlung des Lichtes betreffenden heuristischen Gesichtspunkt," *Annalen der Physik* 17, no. 6 (1905): 132–148. An English translation is at https://en.wikisource.org/wiki/Translation:On_a_Heuristic_Point_of_View_about_the_Creation_and_Conversion_of_Light.

23. For wave-particle duality, see Einstein, "On the Development of Our Views Concerning the Nature and Constitution of Radiation," *Physikalische Zeitschrift* 10 (1909): 817–826; for spontaneous quantum leaps in an electron's energy state, see Einstein, "Emission and Absorption of Radiation in Quantum Theory," *Proceedings of the German Physical Society* 18 (1916): 318–323. An excellent account of these works for the general reader is in Stone, *Einstein and the Quantum*, 136–140 and 184–187. Einstein developed the last idea in collaboration with a young Indian physicist named Satyendra Nath Bose; the new state of matter they proposed eventually came to be known as a "Bose-Einstein condensate." Physicist and science historian A. Douglas Stone, however, argues that "the generosity of the 'Bose-Einstein' designation is not widely appreciated, as few physicists realize that Bose said not a word about the quantum ideal gas in his seminal paper. The paper that does predict quantum condensation belongs to Einstein alone, and it is a masterwork." See Stone, *Einstein and the Quantum*, 237.

24. Stone, *Einstein and the Quantum*, 3.

25. Schrödinger, in one of his seminal papers on wave mechanics in 1926, wrote that his wave equation had been inspired by the "infinitely far-seeing remarks by Einstein." Quoted in Pais, *Subtle Is the Lord*, 439. Likewise, Louis de Broglie explained, "After long reflection in solitude and meditation, I suddenly had the idea, during the year 1923, that the discovery made by Einstein in 1905 should be generalized by extending it to all material particles and notably to electrons." From the preface to de Broglie's reedited PhD thesis of 1924, quoted in Pais, *Subtle Is the Lord*, 436. Max Born also acknowledged Einstein as a key influence in his Nobel Prize–winning work. "An idea of Einstein's gave me the lead," he explained. "He had tried to make the duality of particles—light quanta or photons—and waves comprehensible by interpreting the square of the optical wave amplitudes as probability density for the occurrence of photons. This concept could at once be carried over to the ψ-function: $|\psi|^2$ ought to represent the probability density for electrons (or other particles)." From Born, "The Statistical Interpretation of Quantum Mechanics," 262.

26. In recommending Schrödinger and Heisenberg for the highest honor in all of science, Einstein made his motivations clear: "I am convinced that this theory undoubtedly contains a part of the ultimate truth." He nominated them both in 1928, and again in 1931. Heisenberg won in 1932, after which Einstein recommended Schrödinger for a third time. Schrödinger was finally awarded the Nobel in 1933. See Pais, *Subtle Is the Lord*, 448. Likewise, Louis de Broglie, "with Einstein's strong support, received the Nobel Prize in 1929, a mere four years after his work became widely known." See Stone, *Einstein and the Quantum*, 252. Einstein also nominated Max Born for the Nobel in 1928; see Greenspan, *The End of the Certain World*, 190.

27. Bohr, "Discussion with Einstein," 218.

28. Bohr, "Discussion with Einstein," 201.

29. Einstein, "Physics and Reality" (1936), in *Out of My Later Years*, 92.

30. Einstein, "Physics and Reality," 88.

31. Einstein's words as recalled by Pauli, "Albert Einstein and the Development of Physics" (1958), in *Writings on Physics and Philosophy*, 121.

32. Einstein to Max Born, Sept. 7, 1944, in Born, *The Born-Einstein Letters*, 146.

33. Einstein to Max Born, Dec. 4, 1926, in Born, *The Born-Einstein Letters*, 88.

34. See, for instance, Robert Young, "Scientists Win Physics Nobel Prize for Proving Einstein Wrong," ScienceAlert, Oct. 5, 2022, www.sciencealert.com/scientists -win-physics-nobel-prize-for-proving-einstein-wrong.

35. Einstein to Max Born, Mar. 3, 1947, in Born, *The Born-Einstein Letters*, 155. The 2022 Nobel Prize in Physics was in fact awarded for the demonstration that quantum entanglement is a physical fact. See Young, "Scientists Win Physics Nobel Prize for Proving Einstein Wrong."

36. Einstein, *Out of My Later Years*, 98.

37. Einstein to Max Born, April 1948, in Born, *The Born-Einstein Letters*, 166.

38. Born, *The Born-Einstein Letters*, 166.

39. From a lecture Einstein gave in Berlin in 1927, quoted in Pais, *Subtle Is the Lord*, 443.

40. Einstein to Max Born, Sept. 7, 1944, in Born, *The Born-Einstein Letters*, 146.

41. Einstein and Infeld, *Evolution of Physics*, 75; Einstein to Max Born, Apr. 1948, in Born, *The Born-Einstein Letters*, 170.

42. Einstein to Erwin Schrödinger, Dec. 22, 1950, in Einstein, *Letters on Wave Mechanics*, 44.

43. Einstein, *The Ultimate Quotable Einstein*, 362–363.

44. Einstein and Infeld, *Evolution of Physics*, 292.

45. Einstein, *The Ultimate Quotable Einstein*, 362.

46. For an outstanding overview of these issues aimed at the general reader, I recommend Smolin, *Einstein's Unfinished Revolution*. For a humorous but extremely erudite exploration of these topics, see Peter Woit's brilliant blog *Not Even Wrong* (www .math.columbia.edu/~woit/wordpress/) and his excellent book of the same name. For a more succinct treatment that presupposes a fair degree of familiarity with physics, see George Ellis, "Physics on Edge," *Inference* 3, no. 2 (2017), https://inference-review.com /article/physics-on-edge.

47. Einstein, "The Problem of Space, Ether, and the Field in Physics" (1934), in *Ideas and Opinions*, 310.

48. Einstein and Infeld, *Evolution of Physics*, 152.

49. Einstein and Infeld, *Evolution of Physics*, 213.

50. Einstein, *The Ultimate Quotable Einstein*, 93–94.

51. Einstein, *The Ultimate Quotable Einstein*, 341.

52. Hermanns, *Einstein and the Poet*, 59.

53. Hoffmann and Dukas, *Albert Einstein: Creator and Rebel*, 254.

54. Einstein to Maurice Solovine, written when he was seventy years old, in Goldman, *Einstein's God*, 72–73.

55. Hoffmann and Dukas, *Albert Einstein: Creator and Rebel*, 254.

56. Quoted in Goldman, *Einstein's God*, 74.

57. Quoted in Hoffmann and Dukas, *Albert Einstein: Creator and Rebel*, 257.

58. Hermanns, *Einstein and the Poet*, 13.

59. Hermanns, *Einstein and the Poet*, 109.

60. Einstein, *The Ultimate Quotable Einstein*, 491.

61. Eduard Einstein to Albert Einstein, May 1, 1926, in *CP* 15, doc. 274. See also, in particular, the other letters in this correspondence in *CP* 15 and 16.

62. Eduard Einstein to Albert Einstein, Nov. 14, 1926, *CP* 15, doc. 414.

63. Eduard Einstein to Albert Einstein, Nov. 14, 1926, *CP* 15, doc. 414.

64. Albert Einstein to Eduard Einstein, between Nov. 14 and Dec. 12, 1926, *CP* 15, doc. 415. The remarkable phrase "the highest stage of consciousness as the highest ideal" is a faithful rendition of Einstein's original German ("die höchste Stufe des Bewusstseins als höchstes Ideal").

65. Einstein, "What I Believe."

66. Albert Einstein to Eduard Einstein, between Nov. 14 and Dec. 12, 1926, *CP* 15, doc. 415.

67. Einstein to Hans Mühsam, end of June 1952, in Seelig, *Albert Einstein*, 230–231.

68. Hermanns, *Einstein and the Poet*, 59.

69. Einstein to Hans Mühsam, July 24, 1947, in Seelig, *Albert Einstein*, 231.

70. Einstein, *Ideas and Opinions*, 53.

71. Viereck, *Glimpses of the Great*, 447.

72. Einstein to the queen of Belgium, Mar. 20, 1936, in *The Ultimate Quotable Einstein*, 55.

73. Viereck, *Glimpses of the Great*, 448.

74. Einstein to the queen of Belgium, Sept. 19, 1932, in Einstein, *The Ultimate Quotable Einstein*, 434–435.

75. Albert Einstein to Eduard Einstein, Apr. 17, 1926, in *CP* 15, doc. 257.

76. Einstein, "What I Believe."

77. Einstein, *The Ultimate Quotable Einstein*, 415.

78. Einstein, "What I Believe."

79. From Blake's poem "Auguries of Innocence" (1803), at www.poetryfoundation.org/poems/43650/auguries-of-innocence.

80. Quoted in Moszkowski, *Conversations with Einstein*, 202.

81. Spinoza, *Ethics*, part V, proposition 31, scholium, in *Spinoza: Complete Works*, 377.

82. Spinoza, *Ethics*, part V, preface; part V, proposition 23, scholium, both in *Spinoza: Complete Works*, 363, 374.

83. Schopenhauer, *The World as Will and Representation*, 2:463.

84. Hermanns, *Einstein and the Poet*, 104.

85. Quoted in Goldman, *Einstein's God*, 92–93.

86. Hermanns, *Einstein and the Poet*, 128.

87. Hermanns, *Einstein and the Poet*, 59.

88. Quoted in Goldman, *Einstein's God*, 92.

89. Hermanns, *Einstein and the Poet*, 65.

90. Hermanns, *Einstein and the Poet*, 103.

91. Hermanns, *Einstein and the Poet*, 136.

92. Hermanns, *Einstein and the Poet*, 138, 136.

93. Hermanns, *Einstein and the Poet*, 143. The meeting is also lovingly recalled in Jeremy Patrick Miller's obituary, *Patriot Ledger*, Sept. 17, 2015, at www.patriotledger .com/article/20150917/NEWS/150916043.

Epilogue: Summons to Ascension

1. Actually, the reason it's the *allegory* of the cave is that no one knows exactly what anything in the fable is supposed to symbolize, or whether Plato even intended there to be any single "correct" interpretation. My reading is closely aligned with the vast majority of scholars who've studied Plato through history, and more importantly, it aligns well with the ideas throughout the rest of his writings on the deceptiveness of the senses and the need to rise to an intelligible realm beyond them. But ultimately this is only my own interpretation.

2. *Republic*, 515e–516c, in *Plato: Complete Works*, 1133–1134.

3. *Republic*, 517b, in *Plato: Complete Works*, 1135.

4. *Republic*, 519d–520d, in *Plato: Complete Works*, 1136–1137.

5. Hermanns, *Einstein and the Poet*, 109.

6. Einstein to Marga Planck (Max Planck's widow), Oct. 1947, shortly after Planck passed away. In Fölsing, *Albert Einstein*, 729.

7. Hermanns, *Einstein and the Poet*, 135.

8. From an undated conversation, quoted in Moszkowski, *Conversations with Einstein*, 49.

9. Einstein, "On the 200th Anniversary of Isaac Newton's Death."

10. Einstein, "Isaac Newton" (1942), in *Out of My Later Years*, 219.

11. Hermanns, *Einstein and the Poet*, 26.

12. From Einstein's preface to Planck, *Where Is Science Going?*, 13.

13. Hermanns, *Einstein and the Poet*, 70; Einstein, *Ideas and Opinions*, 41 (emphasis added).

14. Spinoza to Henry Oldenburg, Nov. 20, 1665, in *Spinoza: Complete Works*, 849–850.

15. Schopenhauer, *World as Will and Representation*, 2:18.

16. Hermanns, *Einstein and the Poet*, 94.

17. Hermanns, *Einstein and the Poet*, 61.

18. There has been a wave of Nobel Prizes in recent decades for research that has either confirmed or expanded on Einstein's ideas, including: Bose-Einstein condensates (2001), cosmic expansion (2011), gravitational waves (2017), laser physics (2018), the cosmic background radiation (2019), and black holes (2020). A list of all Nobel Prizes in Physics can be found on the Nobel Foundation website: www.nobelprize.org/prizes /lists/all-nobel-prizes-in-physics/.

19. See Isaacson, *Einstein*, 5.

20. Quoted in Einstein, *Einstein on Peace*, 350. Elsewhere, Einstein asserted that "my participation in the production of the atomic bomb consisted of one single act: I signed a letter to President Roosevelt, in which I emphasized the necessity of conducting large-scale experimentation with regard to the feasibility of producing an atom bomb." From a letter Einstein wrote to the editor of *Kaizo* magazine in Japan, explaining

his views on pacifism and nuclear weapons, dated Sept. 20, 1952; in *Einstein on Politics*, 488. Einstein wasn't just dodging guilt or shirking responsibility. Robert Oppenheimer, who oversaw the actual building of the bomb, agreed that Einstein's role was negligible. "Einstein is often blamed or praised or credited with these miserable bombs," he wrote. "It is not in my opinion true...by 1932 the experimental evidence for the interconvertibility of matter and energy which he had predicted was overwhelming...Einstein himself is really not answerable for all that came later." From a speech given on the tenth anniversary of Einstein's death, in Oppenheimer, "Oppenheimer on Einstein," 38.

21. From Leo Szilard's reminiscences on the day of Einstein's death, April 18, 1955, in Sayen, *Einstein in America*, 151.

22. One of the passages Einstein marked in his German edition of the Tao Te Ching was section 43: "The softest thing on earth overtakes the hardest thing on earth. The non-existent overtakes even that which has no interstices. From this one recognizes the value of non-action. Teaching without words, the value of non-action is attained by but few on earth." English translation from Lao Tzu, *Tao Te Ching*, 47. Einstein also marked verse 63, which likewise discusses *wu wei*. The astute reader might be thinking that this is not what *wu wei* really means, that Taoism does not demand we literally do nothing at all, and that non-action is a much more nuanced concept than this. All true, but all beside the point. The point here is Einstein's obvious familiarity with, and interest in, Taoism. It's doubtful that he understood its subtlest notions in all their depth, but that doesn't matter for our purposes.

23. Plato makes this same point about the divine light initially being blinding when we first emerge from the cave in *Republic*, 515c–516b, in *Plato: Complete Works*, 1133–1134.

24. From a radio broadcast Einstein made on June 22, 1940, after receiving American citizenship, entitled "I Am an American." These comments were made before anyone had even begun building an atomic bomb; Einstein was referring to more conventional weapons of war, which were already destructive enough. But obviously he would have thought the same points applied all the more forcefully to the atomic weapons that appeared on the scene a few years later. See Einstein, *Einstein on Peace*, 312.

25. Einstein, *Einstein on Politics*, 387.

26. Einstein, *Einstein on Peace*, 312.

27. From "Einstein on the Atomic Bomb," interview by Raymond Swing, part 2, *Atlantic Monthly*, Nov. 1947, in *Out of My Later Years*, 199.

28. Hermanns, *Einstein and the Poet*, 143.

29. Hermanns, *Einstein and the Poet*, 109.

30. Viereck, "What Life Means to Einstein," 113.

31. Einstein, *The Ultimate Quotable Einstein*, 237.

32. Einstein, "Science and God," 378.

33. Einstein, *The Ultimate Quotable Einstein*, 239.

34. Einstein, "On the 200th Anniversary of Isaac Newton's Death."

35. Plato, *Phaedrus*, 247c, in *Plato: Complete Works*, 525; *Phaedrus*, 247c–d, trans. Harold N. Fowler, www.perseus.tufts.edu/hopper/text?doc=Perseus%3Atext%3A1999.01.0174%3Atext%3DPhaedrus%3Asection%3D247c.

36. All quotes in this paragraph from Plato, *Phaedrus*, 247c, in *Plato: Complete Works*, 525.

37. Plato, *Letter VII*, 341c, in *Plato: Complete Works*, 1659.

38. Plato, *Letter VII*, 341c, in *Plato: Complete Works*, 1659.

39. From a little poem Einstein wrote in honor of Isaac Newton on the three hundredth anniversary of his birth, in Hoffmann and Dukas, *Albert Einstein: Creator and Rebel*, 141.

INDEX

Index

Index

Index

Index

Index

Index

Index

Tao Te Ching in, 115, 117 (photo), 119,
    120 (photo), 139, 253n8, 277n123,
    277n125, 300n22
Upanishads in, 111–112, 163
life
    meaning of, 46, 131, 201–205
    as sacred, 130–132
    as tapestry, 142
light
    divine, blinding effects of, 214, 300n23
    nature of, 196–197
Lorentz, H. A., 64
love
    intellectual, of God, 167–168
    in mainstream religions, 50, 252n2

Macalister, T., 244n81
Macdonell, Arthur A., 247n42
Mach, Ernst, 44, 64, 175
macrocosm, 70, 89, 160, 163, 169, 211
Magna Graecia, 66, 76, 102, 258n19
magnetism, 53
Mahavira, 133
mainstream religions
    condemnation of cosmic religion by, 42,
        250n76
    curiosity discouraged in, 58–60, 254n60
    decline of, 36
    dualism in, 13, 119–121
    Einstein's dismissal of, 13, 17, 242n59
    emotion in, 49–50, 252–253nn2–4
    ethics in, 152–156
    pacifism in, 131–132
    science as offspring of, 73–74
    as second-phase religions, 30–36, 246n9
    See also specific religions
"Manifesto to the Europeans," 279n22
Marcus, Robert, vii
Maric, Mileva, 241n53, 256n5
Martínez, Alberto, 23
martyrs, 127
mass–energy equivalence. See $E = mc^2$
materialism
    scientific, 155
    transcendental, 114 (photo), 164

mathematics
    origins of, 74, 82
    in revelation of reality, 64, 85, 92
    simplicity in, 64, 89
    Spinoza's use of, 92, 101–102
    in understanding of the divine, 78, 80,
        84–89
    in understanding of harmony, 62, 66,
        70–72, 76, 80–81
matrix mechanics, 195–196, 295n8
matter
    emergence of mind from, 179–180
    vs. field, 144–145
    See also dualism; nondualism
Maxwell, James Clerk, 64, 169–170,
    172
maya, 51–52
McEvilley, Thomas, 68, 260n38
McGrath, Alister, 24, 46, 252n105
meat. See vegetarianism
"mediocre" minds, 45
meditation, 2–3, 68, 180, 248n53
Mein Weltbild (Einstein), 274n78
memory, 69, 150
Menand, Louis, 23
Mercury, 268n168
metaphysical monism, 162–163, 179
metaphysics, 244n77
microcosm, 70, 89, 160, 163, 169, 211
microscopes, 96–97
Middle Ages, 75
Miletus, 65, 257n17
militarism, German, 61, 126, 279n20
mind, human
    in cosmic religion, 6
    curiosity as driver of, 58
    emergence from matter, 179–180
    as eternal, Spinoza on, 161, 206, 212
    evolution of, 7–8
    as identical to the Infinite, 211–212
    limits of, 193–194, 198–201, 205
    "mediocre," 45
    as outside realm of science, 13,
        239n13
    wonder of, 51–58

Index

Index

poetry
 as language of atoms, 195
 of Tagore, 112–113, 215
poisoned arrow parable, 59–60
Polycrates, 66
Porphyry, 257n15, 261n47
portraits, in Einstein's home, 37, 135
   (photo), 136, 248n49, 282n67
Pound, Ezra, 112
prayer, Einstein on, 43, 250n78
priestly caste, 35–36
primates, 57
primordial singularity, 15–16
Principia Mathematica (Newton), 87
Proclus, 76
psychoanalysis, 181
psychology, 178–192
 dream analysis in, 182–183, 187–188,
   291n30
 Jung's work with Pauli in, 182–192,
   292n44
 Pauli's work on synthesis of physics
   and, 181–192
 similarities between physics and, 14,
   162, 180
psycho-physical monism, 184, 189, 192
Pugwash peace conferences, 239n4
pure thought, 52, 63, 64, 89, 187, 195
Pyramid of Unas, 86
Pythagoras, 63–90
 death of, 273n75
 Democritus's book about, 82, 266n143
 education of, 65–66, 257nn15–17,
   262n61
 fame of, 65
 hieros logos and, 237n5
 as inspiration for cosmic religion, 6
 life of, 65–70, 256n7
 reliability of sources on, 256n7
 schools founded by, 66, 68, 258n17
 spiritual practice of, 67–69
 spiritual teachings of, 69–70
Pythagoreans, 63–90
 on astronomy, 66–67, 258n24, 262n66
 dependency thesis and, 73–75, 79

Einstein's introduction to, 70–71
Einstein's writings mentioning, 82–84,
   266n140
on free will, 151–152
fundamental faith of, 70–73,
   84–85, 118
on human body, 67, 259nn26–27
on Infinity, 163
Jainism and, 68, 260n38
on mathematical simplicity, 64, 89
on math in understanding of the divine,
   78, 80, 84–89
on math in understanding of harmony,
   62, 66, 70–72, 76, 80–81
misunderstanding of Einstein's views
   on, 25, 244n82
on numbers, significance of, 66, 71, 76,
   80–81, 84, 89
oral tradition of, 76, 264nn102–103
Pauli influenced by, 72, 80, 190–192
physics influenced by, 71–73
Plato influenced by, 75–78, 102, 191,
   272n52
on reincarnation, 69, 261n47
on simplicity, 72, 81
and Spinoza's pantheism, 100
spiritual practice of, 69–70
stereotypes of, 66
on synthesis of science and spirituality,
   80–81
vegetarianism of, 69, 132, 134, 260n37
See also specific followers
Pythagorean theorem, 70, 92, 266n140

quantum entanglement, 197n35, 198
quantum mechanics
 atom models in, 194–196, 295n8
 current status of, 199–201
 Einstein's opposition to, 194–199
 establishment of basics of, 174
 matrix mechanics in, 195–196, 295n8
 objective reality in, 194–195
 Pauli's contributions to, 175, 289n96
 as theory of everything, 196
 wave function in, 180, 196, 295n8

Index

Index

Upanishads
  ahimsa in, 132, 281n57
  arrival in Western world, 37
  Chandogya, 112, 132, 163, 281n57
  Einstein's reading of, 111–112, 163
  Infinity in, 163
  as inspiration for cosmic religion, 6
  nondualism in, 163
  Path of Knowledge in, 286n38
  Schopenhauer on, 37, 38 (photo), 104,
    110, 167
  unity of consciousness and cosmos
    in, 41

Vahed, G., 240n42
Vaishnavism, 282n73
van den Enden, Franciscus, 93–94
Vedanta, Advaita, 111–113, 166
Vedas
  ahimsa in, 132, 281n55
  Schopenhauer on, 37–39, 38
    (photo), 110
vegetarianism
  of Einstein, 41, 130–131
  in Hinduism, 133
  in Jainism, 133, 136
  misunderstanding of Einstein's views
    on, 41, 130
  of Pythagoreans, 69, 132, 134, 260n37
Venus, 66
visions, religious, 86
visual thinking, 86, 268n164

Wachsmann, Konrad, 248n50
Watson, James, 180, 290n21
wave function, 180, 196, 295n8
Weinberg, Steven, 15, 240n24
Western religions. *See* mainstream
  religions; *specific religions*
Western world
  arrival of Eastern spirituality in, 36–40
  influence of ancient Greece on, 75–79
"What I Believe" (Einstein), 49
"What Is an Agnostic?" (Russell), 44
What Is Life? (Schrödinger), 290n21

White, Michael, 273n71
Whitehead, Alfred North, 74, 75
Wilber, K., *Quantum Questions*, 3–4
Wilhelm, Richard, 115–116, 117 (photo),
  163, 276n123, 277n127
Wilson, E. O., 74
Winteler, Jost, 279n20
wisdom, inner, 184–188
The Wit and Wisdom of Gandhi, 111
wonder, 49–62
  abuse and exploitation of, 57, 58
  in atheism, lack of, 45–47
  as central to cosmic religion, 47, 49–50
  of children, 53
  curiosity combined with, 57–62
  as essential emotion, 49–53
  inspired by Einstein, 55–57
  of the mind, 51–58
  mysteries in, 49–54
  scientific research as source of, 34
  of scientists, 54
  ubiquity of, 51
The World as I See It (Einstein),
  274n78
The World as Will and Representation
  (Schopenhauer), 37, 247n46
world government, 239n5, 243n65
world peace. *See* pacifism
The World's Great Scriptures (Browne),
  111–112, 276n106, 277n123
World War I, pacifist opposition to, 126,
  279n22
World War II
  abuse of wonder in, 57
  bombing of Japan in, 129, 213
  pacifist opposition to, 127–129
Wright, Phyllis, 274n78
wu wei, 213, 300n22

Yahweh, 259n29, 280n49
Yeats, W. B., 112
yin-yang symbol, 121

zeal, religious, 34
Zhmud, Leonid, 267n152

Credit: Fagun Thakrar

Kieran Fox, MD PhD, studied medicine at Stanford University and holds a doctorate in cognitive neuroscience from the University of British Columbia. He is a physician-scientist at the University of California San Francisco, where his research centers on the neural mechanisms and therapeutic potential of meditation practices and psychedelic medicines. He lives in San Francisco. This is his first book.